Routledge Revivals

An Objective Theory of Probability

This reissue of D. A. Gillies highly influential work, first published in 1973, is a philosophical theory of probability which seeks to develop von Mises' views on the subject. In agreement with von Mises, the author regards probability theory as a mathematical science like mechanics or electrodynamics, and probability as an objective, measurable concept like force, mass or charge. On the other hand, Dr Gillies rejects von Mises' definition of probability in terms of limiting frequency and claims that probability should be taken as a primitive or undefined term in accordance with modern axiomatic approaches.

This of course raises the problem of how the abstract calculus of probability should be connected with the 'actual world of experiments'. It is suggested that this link should be established, not by a definition of probability, but by an application of Popper's concept of falsifiability. In addition to formulating his own interesting theory, Dr Gillies gives a detailed criticism of the generally accepted Neyman-Pearson theory of testing, as well as of alternative philosophical approaches to probability theory. The reissue will be of interest both to philosophers with no previous knowledge of probability theory and to mathematicians interested in the foundations of probability theory and statistics.

T0199654

An Objective Theory of Probability

D. A. Gillies

Routledge
Taylor & Francis Group

First published in 1973
by Methuen & Co.

This edition first published in 2011 by Routledge
2 Park Square, Milton Park, Abingdon, Oxon, OX14 4RN

Simultaneously published in the USA and Canada
by Routledge
52 Vanderbilt Avenue, New York, NY 10017

Routledge is an imprint of the Taylor & Francis Group, an informa business

© 1973 D. A. Gillies

Notice:
Product or corporate names may be trademarks or registered trademarks, and are
used only for identification and explanation without intent to infringe.

Publisher's Note
The publisher has gone to great lengths to ensure the quality of this reprint but
points out that some imperfections in the original copies may be apparent.

Disclaimer
The publisher has made every effort to trace copyright holders and welcomes
correspondence from those they have been unable to contact.

ISBN 13: 978-0-415-61792-5 (hbk)
ISBN 13: 978-0-203-82879-3 (ebk)
ISBN 13: 978-0-415-61865-6 (pbk)

An Objective Theory of Probability

D. A. GILLIES

The following account is based on the conception
of probability theory as a special science of the
same sort as geometry or theoretical mechanics.

– von Mises, 1919

The relations between probability and experience
are also still in need of clarification. In investigating
this problem we shall discover what will at first
seem an almost insuperable objection to my
methodological views. For although probability
statements play such a vitally important rôle in
empirical science, they turn out to be in principle
impervious to strict falsification. Yet this very
stumbling block will become a touchstone upon
which to test my theory, in order to find out
what it is worth.

– Popper, 1934

METHUEN & CO LTD
11 New Fetter Lane, London EC4

First published 1973 by
Methuen & Co Ltd

© *1973 D. A. Gillies*

Printed in Great Britain by
William Clowes & Sons Limited,
London, Colchester and Beccles

SBN 416 77350 8

Distributed in the U.S.A. by
HARPER & ROW PUBLISHERS, INC.
BARNES & NOBLE IMPORT DIVISION

Contents

Contents

Preface

The aim of this book is to present a philosophical theory of probability which can best be considered as a development of von Mises' views. I am entirely in agreement with von Mises in regarding probability theory as a mathematical science similar to mechanics or electrodynamics, and probability itself as an objective measurable concept similar to mass or charge. In this sense I am presenting an objective theory of probability – as opposed to subjective or logical accounts.

The point where I differ from von Mises is often, though wrongly in my opinion, taken as the essential feature of his theory: namely the definition of probability as limiting frequency. This definition, as is well-known, leads to mathematical complexities, and, generally speaking, does not harmonize at all well with the measure-theoretic approach to probability theory introduced by Kolmogorov and almost universally accepted among mathematicians. At the same time it is often felt that some kind of definition of probability in terms of frequency is needed in order to establish the connection between the abstract calculus of probability and the world of actual experiments. My suggested resolution of this difficulty is that the link between theory and experience should be established *not* by a definition but rather by an application of Popper's concept of falsifiability.

In the Introduction I give a critical account of the various current standpoints in the philosophy of probability. Against this background, it is possible to give a more detailed sketch of my own theory, and to explain where it stands in the spectrum of competing views. The remainder of the book is then devoted to developing the theory in detail. Throughout I make free use of the standard terminology of probability theory, but for the benefit of the reader unacquainted with the mathematical theory an appendix is included which explains the meanings of all the technical terms. Any such reader who goes through this appendix first, and then, if necessary, refers back to the technical

x   *Preface*

terms when they occur, should have no difficulty in following the text. In this way I hope the book will prove interesting and comprehensible both to mathematicians interested in the foundations of probability and statistics, and to philosophers who have not studied the mathematical theory.

Extracts from the Introduction and Part III have already appeared in the *British Journal for the Philosophy of Science*; and from Part I in *Synthese*. I would like to thank the publishers of these journals, that is the Cambridge University Press and the D. Reidel Publishing Company, for permission to reprint these passages.

It remains only to express my thanks to the many friends and colleagues who have helped me in writing this book. Much of the book is a revised version of parts of my Ph.D. thesis, and my greatest debt of gratitude is thus to my supervisor Imre Lakatos for all his help and encouragement, as well as for introducing me to the Popperian standpoint in philosophy which has very deeply influenced the whole work. Previous versions of many parts of the book were circulated among my friends or read at various seminars in London and Cambridge and I would like to thank all those who offered very valuable comments and criticisms, and especially to mention in this connection Allan Birnbaum, Richard Braithwaite, Jon Dorling, Bruno de Finetti, Ian Hacking, Mary Hesse, Mark Hill, Colin Howson, John Lucas, Hugh Mellor, Sir Karl Popper, Heinz Post, Quentin Rappoport, Alan Stuart, Aidan Sudbury, John Watkins, John Worrall and Elie Zahar. Some particular debts are mentioned in footnotes. The book was mainly written while I was holding a research fellowship at King's College, Cambridge, and I should like to thank all those who made my stay there a most agreeable one.

Chelsea College,   D. A. GILLIES
University of London,
Manresa Road, S.W.3.
January 1973

The Two Traditions in the Philosophy of Probability. Outline of the Present Work

There are two traditions in the philosophy of probability. We shall call them the *scientific* and the *logical*, using these words in a sense which will be defined more precisely later. Of the two the logical is the older. It was first adumbrated by Leibniz at a time when only the most rudimentary beginnings had been made with the mathematical theory. The scientific approach is of much more recent origin. It was first developed in the middle of the nineteenth century by the Cambridge school of Ellis and Venn and can be considered as a 'British empiricist' reaction against the 'Continental rationalism' of Laplace and his followers. It is not my purpose, however, to trace the early history of the subject in detail. I shall begin rather with the years 1919 and 1921 which saw the publication of an important work in each of the two traditions: in the scientific tradition von Mises' *Grundlagen der Wahrscheinlichkeitsrechnung* (1919) and in the logical Keynes' *Treatise on Probability* (1921). Needless to say, both these authors drew heavily on the work of their predecessors; but they both worked out their respective approaches in greater detail than had previously been attempted, and, in particular, both tried to develop the formal calculus of probabilities from their respective philosophical standpoints. I will begin by summarizing their ideas; a procedure which will throw into sharp relief the problems to be dealt with in what follows.

(i) Von Mises' theory of probability

Keynes did not return to the subject of probability after the publication of his treatise. Von Mises, however, extended his ideas in a number of subsequent writings. His *Probability,*

Statistics and Truth (1928) gave a fuller philosophical account of the matter, while his *Wahrscheinlichkeitsrechnung* (1931) provided greater mathematical detail. Finally, in his posthumous work, *Mathematical Theory of Probability and Statistics* (1963), assembled from his papers by his widow Hilda Geiringer, he made a number of mathematical changes and tried to counter some of the criticisms made of his original theory. In Chapter 4 some of these changes will be described. The present summary will deal only with those features of the theory which remained constant throughout this evolution.

In the preface to the third German edition of *Probability, Statistics and Truth* von Mises characterizes his theory thus:

> The essentially new idea which appeared about 1919 (though it was to a certain extent anticipated by A. A. Cournot in France, John Venn in England, and Georg Helm in Germany) was to consider the theory of probability as a science of the same order as geometry or theoretical mechanics.

Concerning this putative science of probability we might first ask: 'what is its subject matter?'. Von Mises answers as follows (1928, p. vii of 1950 preface):

> ...just as the subject matter of geometry is the study of space phenomena, so probability theory deals with mass phenomena and repetitive events.

And again (1928, p. 11):

> The rational concept of probability, which is the only basis of probability calculus, applies only to problems in which either the same event repeats itself again and again or a great number of events are involved at the same time.

The examples he gives of 'mass phenomena' can be divided into three categories. First come 'games of chance' where we deal, for example, with a long sequence of tosses of a particular coin. Second we have certain 'life' or more generally 'biological statistics'. Here we might deal with the set of Germans who were forty years old in 1928 or with the set of plants grown in a certain field. Lastly we have a number of situations which occur in physics, for example, the consideration of the molecules of a particular mass of gas. In all the examples cited a certain 'attribute' occurs at each of the 'events' which make up the mass phenomenon, but this attribute varies from one event to another. For example, on each toss of the coin 'heads' or 'tails' occurs, each of the Germans either dies before reaching the age of forty-one or survives into his forty-second year, the plants in the field yield a certain quantity of pollen, and finally each of the molecules of the gas has a certain velocity. Associated with each mass phenomenon we have a set of attributes which we regard as *a priori* possible. These form the so-called 'attribute space'. Von Mises introduces the technical term 'collective' to describe mass phenomena of the above types. More precisely, he says (1928, p. 12) that a collective 'denotes a sequence of uniform events or processes which differ by certain observable attributes, say colours, numbers, or anything else'.

Corresponding to these sets of 'uniform events' we introduce in our mathematical theory infinite sequences of *elements* $(\omega_1, \omega_2, \ldots)$ which all belong to some abstract attribute space Ω. There is an ambiguity in von Mises' formulations here, as he uses the word 'collective' to denote both these mathematical infinite sequences, and the (necessarily finite) sets of real events to which they are supposed to correspond. I shall in future use the word 'collective' to designate only the mathematical sequences, and will refer to the corresponding sets of events in the real world as 'empirical collectives'. This distinction raises a problem. We have in reality a large but finite set of events and we abstract from this in our mathematical theory to obtain an infinite sequence. But is this procedure of abstraction a valid one? I shall return to this question from time to time in what follows.

Von Mises was an empiricist and he took the theory of probability to be founded on certain empirical laws which are observed to hold for empirical collectives. There are two of these. The first is introduced thus (1928, p. 12):

> It is essential for the theory of probability that experience has shown that in the game of dice, as in all the other mass phenomena which we have mentioned, the relative frequencies of certain attributes become more and more stable as the number of observations is increased.

The relative frequency mentioned here is defined as follows. Suppose we have made n observations and attribute A occurs

$m(A)$ times, then its relative frequency is $m(A)/n$. Von Mises refers to this increasing stability of relative frequencies as (1928, p. 14) 'the primary phenomenon (*Urphänomen*) of the theory of probability'. We, however, will call it the *Law of Stability of Statistical Frequencies*. It is described more precisely by von Mises as follows (1928, p. 14):

If the relative frequency of heads is calculated accurately to the first decimal place, it would not be difficult to attain constancy in this first approximation. In fact, perhaps after some 500 games, this first approximation will reach the value of 0·5 and will not change afterwards. It will take us much longer to arrive at a constant value for the second approximation calculated to two decimal places.... Perhaps more than 10,000 casts will be required to show that now the second figure also ceases to change and remains equal to 0, so that the relative frequency remains constantly 0·50.

This first law of empirical collectives was fairly well known before von Mises. The second law is, however, original to him. Indeed he considers its formulation to be one of his major advances. Speaking of the efforts of his predecessors in the scientific tradition (Venn and others), he says (1928, p. 22): 'These attempts...did not lead, and could not lead, to a complete theory of probability, because they failed to realize one decisive feature of a collective....' This feature of the empirical collective is its lack of order, that is its *randomness*.

Von Mises' treatment of the question of randomness is most interesting. He points out that over a long period of time many different gambling systems have been tried out. We could cite 'Bet on red after a run of three blacks' or 'Bet on every seventh go' as typical examples. However (1928, p. 25): 'The authors of such systems have all, sooner or later, had the sad experience of finding out that no system is able to improve their chances of winning in the long run i.e. to affect the relative frequencies with which different colours or numbers appear in a sequence selected from the total sequence of the game.' In other words, not only do the relative frequencies stabilize around certain values but these values remain the same if we choose, according to some rule, a subsequence of our original (finite) sequence. I shall call this second empirical law the *Law of Excluded Gambling Systems*.

Von Mises now makes a most suggestive comparison (1928, p. 25):

An analogy presents itself at this point which I shall briefly discuss. The system fanatics of Monte Carlo show an obvious likeness to another class of 'inventors' whose useless labour we have been accustomed to consider with a certain compassion, namely, the ancient and undying family of constructors of 'perpetual-motion' machines.

The failure of all attempts to construct a perpetual-motion machine provided excellent evidence for the Law of Conservation of Energy. Conversely we could describe the Law of Conservation of Energy as the Law of Excluded Perpetual-Motion Machines. In just the same way the failure of gambling systems provides excellent evidence for our empirical law of randomness.

So now we have two empirical laws governing the behaviour of empirical collectives. Corresponding to these we formulate two axioms which our mathematical concept of collective must obey. The first – the axiom of convergence – can be stated thus: Let A be an arbitrary attribute of a collective $\mathfrak{C}$, then $\lim_{n\to\infty} m(A)/n$ exists. We now *define* the probability of A in $\mathfrak{C}[p(A,\mathfrak{C})]$ as $\lim_{n\to\infty} m(A)/n$. This is the famous limiting frequency definition of probability.

It has often been objected (e.g. by Keynes) against theories which adopt this definition of probability that they restrict the use of the concept too much. After all, it is claimed, we often speak of probabilities where the idea of limiting frequency is not at all applicable. This alleged disadvantage is, however, considered by von Mises to be a strong point in favour of his theory. He states clearly (1928, p. 9) that 'Our probability theory has nothing to do with questions such as: "Is there a probability of Germany being at some time in the future involved in a war with Liberia?"' We can only, he claims, introduce probabilities in a mathematical or quantitative sense where there is a large set of uniform events, and he urges us to observe his maxim (1928, p. 18): 'First the collective – then the probability.'

There is indeed much to be said for von Mises' desire to limit the scope of the mathematical theory. The history of probability theory affords some curious examples of 'numerical' probabilities. Todhunter (1865, pp. 408–9), for example, records the following evaluations carried out by the eighteenth-century probabilist Condorcet (who was later guillotined). The probability that the whole duration of the reigns of the seven kings of Rome was 257 years was reckoned by him to be 0·000792, while the probability that it was 140 years came to 0·008887. He also calculated the probability that the augur Accius Naevius cut a stone with a razor. This came to the more rounded figure of 10^{-6}.

Von Mises' view on this point is connected with some general theories of his about the evolution of science. On p. 1 (1928) he quotes with approval Lichtenberg's maxim that 'All our philosophy is a correction of the common usage of words' (a maxim which should perhaps be interpreted normatively as it is hardly true empirically of Anglo-Saxon philosophy today). We can, according to von Mises, *start* with the imprecise concepts of ordinary language but when we are constructing a scientific theory we must replace these by more precise concepts. Further, he thinks that these precise concepts should be introduced *by means of explicit definitions*. This shall be called *von Mises' definitional thesis*. The example he cites in this context is the mechanical concept of work. Of course we use the word 'work' in a variety of ways in ordinary language, but in mechanics we define work as force times distance or more precisely we set

$$W_a^b = \int_a^b \mathbf{F} . \, \mathrm{d}s$$

where W_a^b is the work done in moving from a to b in a conservative force field $\mathbf{F}(\mathbf{x})$. Many things ordinarily counted as work are excluded by this definition, e.g. working a typewriter, the work involved in writing a book, the hard work of two teams in a tug-of-war when the rope isn't moving.... The vague concept of ordinary language has been delimited and made more precise by a definition.

This example seems to be entirely fair and shows that some concepts are introduced in accordance with von Mises' definitional thesis. But can all the concepts of exact science be introduced in this fashion? Would this not lead to an infinite regress? I shall examine this matter in the first few chapters of Part I.

So much for the Axiom of Convergence. The second axiom, the Axiom of Randomness, corresponds to the law of excluded gambling systems. It states roughly that the limiting frequency of any attribute in a collective remains the same in any subsequence obtained from the original sequence by a place selection. The difficulties involved in a more precise statement of this axiom will be discussed in Part II below. From these two axioms the mathematical theory of probability follows, or rather *ought* to follow for, as a matter of fact, von Mises had to introduce a third axiom into some of the later versions of his theory. The task of the mathematical theory can be fairly precisely defined. In mechanics, given a system of certain masses with certain initial velocities and acted on by certain forces, we have to calculate, using general laws, the state of the system at a later stage. So similarly in the case of probability (1928, p. 32):

> The exclusive purpose of this theory is to determine from the given probabilities in a number of initial collectives, the probabilities in a new collective derived from the initial ones.

Finally we must add a word about von Mises' attitude to statistics. The general position of statistics suggested by his theory is perhaps something like this. In applied mechanics we study what mechanical theories are appropriate in certain situations and how, in such situations, we can measure the forces and masses involved. Similarly in applied probability theory or statistics we study what probabilistic theories are appropriate in certain situations and how in these situations we can measure the probabilities involved. I think that von Mises would have agreed to this general formulation; but I shall not give a detailed account of his views on statistics, partly because I believe that they are inconsistent with his general view of probability.

(ii) Keynes' theory of probability

When we now turn to Keynes' theory, the whole picture changes. To begin with probability theory is conceived as a branch of logic, and probability itself as a logical relation which holds between two propositions. To explain this let us consider two propositions e_1 and e_2. By the ordinary laws of logic e_1 entails e_1 or e_2, but e_1 does not entail e_1 and e_2. On the other hand we might feel that e_1 partially entails e_1 and e_2 because it entails

2

one half of the conjunct. Suppose such a relation of partial entailment really existed and further we could measure its degree – p say where $0 \leqslant p \leqslant 1$ – then p, for Keynes, would be the probability of e_1 and e_2 given e_1. As he puts it (1921, p. 5):

> We are claiming, in fact, to cognise correctly a logical connection between one set of propositions which we call our evidence and which we suppose ourselves to know, and another set which we call our conclusions, and to which we attach more or less weight according to the grounds supplied by the first.... It is not straining the use of words to speak of this as the relation of probability.

An important feature of this account is that it makes all probabilities conditional. We cannot speak simply of the probability of a hypothesis, but only of its probability relative to some evidence which partially entails it. Keynes remarks (1921, p. 7):

> No proposition is in itself either probable or improbable, just as no place can be intrinsically distant; and the probability of the same statement varies with the evidence presented, which is, as it were, its origin of reference.

At first this would seem to conflict with our ordinary use of the probability concept for we do often speak simply of the probability of some outcome. Keynes would reply that in such cases a standard body of evidence is assumed. Von Mises would agree with Keynes that every probability is conditional, but for him they are conditional on the collectives within which they are defined.

So far we have described the probability relation as 'degree of partial entailment'. But Keynes gives another account of it as follows (1921, p. 4):

> Let our premises consist of any set of propositions h, and our conclusions consist of any set of propositions a, then, if a knowledge of h justifies a rational belief in a of degree α, we say that there is a *probability-relation* of degree α between a and h.

Here Keynes makes the assumption that if h partially entails a to degree α, then given h it is rational to believe a to degree α. Put less precisely he identifies 'degrees of partial entailment' and 'degrees of rational belief'. This assumption seems at first sight highly plausible, but as we shall see later it has been challenged.

The next question we might ask is: 'How do we obtain knowledge about this logical relation of probability, and, in particular, how are the axioms of probability theory to be established from this point of view?' On the general problem of knowledge Keynes adopted a Russellian position. Russell held that some of our knowledge is obtained directly or 'by acquaintance'. His views on what we could know in this way varied, but the set always included our immediate sense perceptions. The rest of our knowledge is 'knowledge by description' and is ultimately based on our 'knowledge by acquaintance'. In analysing the relations between the two sorts of knowledge, Russell thought that his theory of descriptions could play an important rôle. In Russellian vein Keynes writes (1921, p. 14): 'About our own existence our own sense-data, some logical ideas, and some logical relations, it is usually agreed that we have direct knowledge.' In particular we get to know certain probability relations by 'direct acquaintance' or 'immediate logical intuition'. For in the case of some propositions at least, Keynes thinks that (1921, p. 13):

We pass from a knowledge of the proposition a to a knowledge about the proposition b by perceiving a logical relation between them. *With this logical relation we have direct acquaintance.* [My italics]

Though he does admit (1921, p. 15) that 'Some men...may have a greater power of logical intuition than others.'

So we are given a large number of instances of the probability relation by logical intuition. The purpose of an axiom system for probability is largely an economical one: to reduce these many directly cognized instances to a few cases from which the rest will follow. Keynes also thinks that a formal system of probability will include ordinary demonstrative logic as a special case. The object of constructing an axiom system is, he says (1921, p. 133):

...to show that all the usually assumed conclusions in the fundamental logic of inference and probability follow rigorously from a few axioms.... This body of axioms and theorems corresponds, I think, to what logicians have termed the *Laws of Thought*, when they have meant by this something narrower than the whole system of formal truth. But it goes beyond what has been usual, in dealing at the same time with the laws of probable as well as necessary inference.

One point on which Keynes and von Mises agree must now be emphasized. Keynes does not believe that all probabilities have a numerical value. On the contrary, certain probabilities may not even be comparable. As he says (1921, p. 27):

> ...no exercise of the practical judgment is possible by which a numerical value can actually be given to the probability of every argument. So far from our being able to measure them, it is not even clear that we are always able to place them in an order of magnitude. Nor has any theoretical rule for their evaluation ever been suggested.

So if we have two probabilities a variety of situations can hold. They may both have a numerical value. Again, though we may not be able to assign a numerical value to both of them, we will perhaps be able to say that one is greater than the other. In still other cases we may not be able to make any comparison. As Keynes puts it (1921, p. 34):

> I maintain, then, in what follows, that there are some pairs of probabilities between the members of which *no* comparison of magnitude is possible; that we can say, nevertheless, of some pairs of relations of probability that the one is greater and the other less, although it is not possible to measure the difference between them; and that in a very special type of case, to be dealt with later, a meaning can be given to a *numerical* comparison of magnitude.

The set of probabilities is thus not linearly ordered. It has, however, a special kind of partial ordering which Keynes describes on pp. 38–40. I will not go into this in detail here.

What, then, are the cases in which numerical values can be assigned to probabilities? Keynes answers unequivocally (1921, p. 41): 'In order that numerical measurement may be possible, we must be given a number of *equally* probable alternatives.' He even claims that this is something on which most probabilists agree (1921, p. 65): 'It has always been agreed that a numerical measure can actually be obtained in those cases only in which a reduction to a set of exclusive and exhaustive *equiprobable* alternatives is practicable.' So in order to get numerical probabilities we have to be able to judge that a certain number of cases are equally probable and to enable us to make this judgment we need a certain *a priori* principle. This *a priori* principle is called by Keynes the *Principle of Indifference*. The name is original to him but the principle itself, he says, was introduced by James Bernoulli under the name of the *Principle of Non-Sufficient Reason*. Keynes gives the following preliminary statement of the principle (1921, p. 42):

> The Principle of Indifference asserts that if there is no *known* reason for predicating of our subject one rather than another of several alternatives, then relatively to such knowledge the assertions of each of these alternatives have an equal probability.

The trouble with the principle of indifference is that it leads us at once into a number of grave contradictions. These contradictions were discovered by a number of authors, notably by Bertrand and Borel. It is greatly to Keynes' credit that although he advocates the principle of indifference, he gives the best statement in the literature of the objections to it. I will confine myself to stating two of the simplest paradoxes. Consider a book whose colour we don't know. We have no more reason to suppose it is red than that it is not red. Thus using the principle of indifference we have prob(red) = $\frac{1}{2}$. Similarly however prob(blue), prob(green) and prob(black) are all $\frac{1}{2}$, which is absurd. Our second example belongs to the class of paradoxes of 'geometrical' probability. Suppose we have a mixture of wine and water and we know that at most there is 3 times as much of one as of the other, but we know nothing more about the mixture. We have

$$1/3 \leqslant \text{wine/water} \leqslant 3$$

and by the principle of indifference the ratio wine/water has a uniform probability density in the interval (1/3, 3), therefore

$$\text{prob (wine/water} \leqslant 2) = (2 - 1/3)(3 - 1/3) = 5/8.$$

But also

$$1/3 \leqslant \text{water/wine} \leqslant 3$$

and by the principle of indifference the ratio water/wine has a uniform probability density in the interval $(1/3, 3)$, therefore

$$\text{prob (water/wine} \geqslant \tfrac{1}{2}) = (3 - \tfrac{1}{2})/(3 - 1/3) = 15/16.$$

But the events 'wine/water $\leqslant 2$' and 'water/wine $\geqslant \tfrac{1}{2}$' are the same and the principle of indifference has given them different probabilities.

Let us now turn to Keynes' attempted resolution of these difficulties. He believes we can avoid them if we apply the principle of indifference only to cases where the alternatives are finite in number and 'indivisible' (1921, p. 60):

> Let the alternatives, the equiprobability of which we seek to establish by means of the Principle of Indifference, be $\phi(a_1)$, $\phi(a_2), \ldots, \phi(a_r)$, and let the evidence be h. Then it is a necessary condition for the application of the principle, that there should be, relatively to the evidence, *indivisible* alternatives of the form $\phi(x)$.

I will not give a formal definition of this notion of 'indivisibility' but attempt a rough explanation. Suppose we have a group of alternatives and can separate from one of them a subalternative of the same form as one of the original alternatives, then that alternative is divisible. Thus, for example, consider the alternatives red and non-red. We can divide non-red into blue and non-(red or blue) and blue is of the same form as red. Thus this alternative is divisible.

The trouble with this suggestion is that in the case of geometrical probabilities we have an infinite number of alternatives; or, to look at it another way, given any finite number of alternatives we can always divide them into further subalternatives of the same form. For example, consider the interval $(1/3, 3)$ in the wine-water example. If we divide it into n equal subintervals $I_1, \ldots, I_n$, these can always be divided into further subintervals. Keynes attempts to deal with the geometrical case by specifying that we only consider a finite number of alternatives. He does not seem to notice the contradiction with the principle of indivisibility which we have just noted. To bolster his suggestion he points out that we can make the lengths of the subintervals considered as small as we like, and claims that as long as the form of the finite alternatives is specified we shall avoid contradictory conclusions.

On this part of Keynes' work von Mises, in a later edition of *Probability, Statistics and Truth*, made the following comment (1928, p. 75):

> Keynes makes every effort to avoid this dangerous consequence of the subjective theory [i.e. the book paradox described above] but with little success. He gives a formal rule precluding the application of the Principle of Indifference to such a case, but he makes no suggestion as to what is to replace it.

This is unfair, because Keynes does make a suggestion about how to set up a modified principle of indifference. On the other hand the success of his attempt is highly dubious. We have already noted that the treatment of the 'geometrical' case contradicts the principle of indivisibility. It must now be added that it does not appear to avoid the paradox. Suppose in the wine-water example we divide the interval $(1/3, 3)$ into n equal subintervals $I_1, \ldots, I_n$ and consider the event E that there is less than twice as much wine as water. By taking the length of I_i sufficiently small and representing E first as a combination of events of the form wine/water $\in I_j$ and then of events of the form water/wine $\in I_k$, we obtain by suitably modifying the previous argument two different probabilities for E.

Finally we shall give a brief account of Keynes' attitude to statistics. He divides the theory of statistics into two parts (1921, p. 327):

> The first function of the theory is purely *descriptive*. It devises numerical and diagrammatic methods by which certain salient characteristics of large groups of phenomena can be briefly described.... The second function of the theory is *inductive*. It seeks to extend its description of certain characteristics of observed events to the corresponding characteristics of other events which have not been observed. This part of the subject may be called the Theory of Statistical Inference; and it is this which is closely bound up with the theory of probability.

Thus, for example, we might describe the relationship between two finite sets of characteristics by calculating their correlation coefficient. If these sets are samples from a population we then make a statistical inference from the value of the correlation coefficient in the sample to its value in the population. Of course such statistical inferences are just special cases of inductive inferences, i.e. inferences from finite evidence to generalizations based on such evidence. The theory of probability is concerned with the probability that such inductive generalizations receive from the evidence and thus *a fortiori* with a statistical inference. Presumably though the special features of the situation enable the probabilities involved to be calculated more precisely.

(iii) Comparison between von Mises and Keynes

The remarkable thing here is the *point for point disagreement* which exists between the two theories. For von Mises probability theory is a branch of empirical science; for Keynes it is an extension of deductive logic. Von Mises defined probability as limiting frequency; Keynes as degree of rational belief. For one the axioms of probability are obtained by abstraction from two empirical laws; for the other they are perceived by direct logical intuition. It is true that on one point there is some agreement. Neither thinks that all probabilities have a numerical value, but the attitude of the two authors to this situation is very different. For von Mises only probabilities defined within an empirical collective can be evaluated and only these probabilities have any scientific interest. The remaining uses of probability are examples of a crude pre-scientific concept towards which he takes a dismissive attitude. For Keynes on the other hand all probabilities are essentially on a par. They all obey the same formal rules and play the same rôle in our thinking. Certain special features of the situation allow us to assign numerical values in some cases, though not in general. Finally the position of statistics is different in the two accounts. For von Mises it is a study of how to apply probability theory in practice, similar to applied mechanics. For Keynes statistical inference is a special kind of inductive inference and statistics is a branch of the theory of induction.

It is worth noting that this complete *disagreement* on all the philosophical issues is accompanied by complete *agreement* on the mathematical side. Both authors in their books derive the Binomial theorem, the Laws of Large Numbers, the Central

Limit theorem; that is to say all the main results of the mathematical calculus. What is still more surprising is that even after a further fifty years of mathematical developments philosophical differences of a closely related kind still persist. This is the problem which lies behind much of the following investigations.

(iv) Development of the logical tradition after Keynes

Let us now sketch very briefly the developments in the logical tradition after Keynes. The first interesting innovation is due to Ramsey in *Truth and Probability*, 1926. Keynes had taken probability as degree of rational belief. Ramsey altered this to: 'degree of belief subject to certain weak rationality constraints'. He further suggested that we measure degrees of belief by considering the quotients which people would adopt if forced to bet on the events in question. The 'weak rationality constraints' arise because we must arrange our betting quotients so that a Dutch book cannot be made against us, i.e. so that a cunning opponent cannot ensure that we always lose. From this so called 'coherence' condition Ramsey obtained the axioms of probability. One interesting feature of his account is that he thought the bets would be unsatisfactory measures if they were made in money. He therefore developed simultaneously a theory of value and of probability (1926, pp. 176–82). He remarks (p. 179) that his definition of degrees of belief 'amounts roughly to defining the degree of belief in *p* by the odds at which the subject would bet on *p*, the bet being conducted in terms of *differences of value* as defined'. [My italics]

The betting approach was adopted later but independently by de Finetti in his article 'Foresight: Its Logical Laws, Its Subjective Sources' (1937). De Finetti, however, considers money bets rather than value bets. He goes beyond Ramsey in discussing an important problem which arises on any 'degree of belief' account of probability. The account works well in cases like gambling on horses where there is genuinely interpersonal variation in the probabilities assigned. But what about cases in games of chance or in physics where there are apparently objectively measurable probabilities on whose value everyone agrees? How is the existence of such probabilities to be explained on the personalistic 'degree of belief' point of view? De Finetti puts the point like this (1937, p. 152):

It would not be difficult to admit that the subjectivistic explanation is the only one applicable in the case of practical predictions (sporting results, meteorological facts, political events, etc.) which are not ordinarily placed in the framework of the theory of probability, even in its broadest interpretation. On the other hand it will be more difficult to agree that this same explanation actually supplies rationale for the more scientific and profound value that is attributed to the notion of probability in certain classical domains.

To discuss this problem let us take a particular example. We will first describe this from the objective point of view and then examine how de Finetti proposes to treat it from his 'degree of belief' position. The example is this. We have two men M and N. N has a coin which looks fair but for which in fact prob (heads) = 2/3, and not $\frac{1}{2}$. N tosses the coin a large number of times, and M observes the results. M initially supposes that the coin is fair, i.e. prob (heads) = $\frac{1}{2}$. However, after a while he observes that there has been an excessive preponderance of heads. He therefore regards his initial hypothesis prob (heads) = $\frac{1}{2}$ as falsified, and adopts the hypothesis prob (heads) = 2/3 instead. This agrees with his subsequent observations and he finally concludes that prob (heads) really is 2/3 or at least that this hypothesis is well-corroborated. So much for the objective description of this example. We must now examine how de Finetti would deal with it.

On his approach M has to consider at what odds he would bet if he was forced to do so by N. He decides that his initial betting quotient on heads should be $\frac{1}{2}$. He also decides *a priori* what conditional bets he would make on future tosses. A conditional bet on B given A at rate p is defined thus. If A occurs we bet on B at rate p. If A does not occur, we call off the bet. In the particular case we are considering, M has to propose conditional betting quotients on the result E_{n+1} say of the $n + 1$th toss given all possible results of the first n tosses. These various betting quotients must satisfy the further condition of exchangeability. In our present case this means that the *a priori* betting quotient for getting a particular sequence of r results (π_r say) on tosses $i_1, i_2, \ldots, i_r$ depends perhaps on r but not on $i_1, \ldots, i_r$ for any $r \geqslant 1$. Intuitively this amounts to saying that we regard any

particular group of r tosses as 'exchangeable' with any other –
an apparently reasonable requirement.

The various betting quotients introduced must of course be
coherent. Hence they must satisfy the axioms of probability
and in particular Bayes' theorem. Therefore using fairly standard
methods of reasoning de Finetti shows that whatever (within
broad limits of course) a priori probability $p(E_{n+1})$ we introduce
for getting say heads on the $n + 1$th toss, the conditional
probability $p(E_{n+1}|A)$ will tend to r/n where A states that there
were r heads on the first n tosses. De Finetti assumes that if we
observe A we will bet on E_{n+1} at the betting quotient $p(E_{n+1}|A)$.
Thus whatever our *a priori* betting quotients, our *a posteriori*
betting quotients will tend more and more to the observed
frequencies. The objectivist will incorrectly suppose that we are
discovering the 'true' probability. But, according to de Finetti,
all that is happening is this. Two men starting with different
personal degrees of belief are gradually forced by the constraints
of coherence and exchangeability to adopt degrees of belief
which become increasingly similar and approximate more and
more closely to the observed frequency.

It is worth stressing two assumptions on which the above
argument depends. First of all a certain *a priori* distribution
must be adopted before the first toss, and this distribution is
never given up but rather replaced by an *a posteriori* distribution. That is to say we can never state that our *a priori* probabilities were wrong and we should instead have adopted such and
such different *a priori* probabilities. Secondly it is assumed that
when we bet on E_{n+1} having observed result A on the first n
tosses, we use our original conditional betting quotient
$p(E_{n+1}|A)$. De Finetti himself lays emphasis on these two points
in the following passage (1937, p. 146):

Whatever be the influence of observation on predictions of
the future, it never implies and never signifies that we *correct*
the primitive evaluation of the probability $p(E_{n+1})$ after it
has been *disproved* by experience and substitute for it another
$p^*(E_{n+1})$ which *conforms* to that experience and is therefore
probably *closer to the real probability*; on the contrary, it
manifests itself solely in the sense that when experience
teaches us the result A on the first n trials, our judgment will

be expressed by the probability $p(E_{n+1}|A)$, i.e. that which
our original opinion would already attribute to the event
E_{n+1} considered as conditioned on the outcome A. Nothing
of this initial opinion is repudiated or corrected; it is not the
function p which has been modified (replaced by another p^*),
but rather the argument E_{n+1} which has been replaced by
$E_{n+1}|A$, and this is just to remain faithful to our original
opinion (as manifested in the choice of the function p) and
coherent in our judgment that our predictions vary when a
change takes place in the known circumstances.

In our criticism of de Finetti's argument we will attack
precisely these two assumptions. But before doing so, it is worth
pointing out that the argument, if it were correct, would avoid
the difficulties concerned with the Principle of Indifference
which cropped up in Keynes' theory. Keynes held that we
could only assign numerical values where *a priori* judgments of
equiprobability based on the principle of indifference were
possible. Further, these *a priori* judgments had to be such as to
command the assent of any man capable of logical intuition.
On de Finetti's account a wide variety of *a priori* judgments
are possible – this is the subjective side of the theory – but once
these *a priori* judgments have been made, logical constraints
force the rational man to alter his probabilities in accordance
with Bayes' theorem till eventually, after enough evidence has
been collected, agreement in judgment is reached.

The criticism I shall now give of de Finetti's argument was
first clearly formulated by Hacking (1967, §2, 3), though there
are hints of it (as Hacking points out) in Suppes (1966, §4).
Hacking attacks the second of the assumptions which we listed
above. He observes that there is no need, so far as coherence and
exchangeability are concerned, to adopt the conditional betting
quotient $p(E_{n+1}|A)$ as one's betting rate on E_{n+1} once A has
actually occurred. All the initial betting quotients $p(E_1)$,
$p(E_{n+1})$, $p(E_{n+1}|A)$ etc. refer to actual or hypothetical bets made
before the result of the first toss is observed. Coherence and
exchangeability require that they should obey certain conditions. But suppose one or more results are known. We are now
in a different betting situation and the previous constraints no
longer apply. In particular, if the result A of the first n tosses

is known, we could coherently adopt any betting quotient (p say) on E_{n+1} provided we adopt $1 - p$ on $\bar{E}_{n+1}$. We do not have to put $p = p(E_{n+1}|A)$.

Once this is pointed out, we can easily find situations in which someone might quite reasonably not want to set $p = p(E_{n+1}|A)$. Suppose again that N is forcing M to bet. N shows M, from a distance, a coin which looks a little bent. He tells him that the coin is biased, but refuses to say in which direction it is biased. M has to make the usual conditional and absolute bets. As the game proceeds he is told the results of the tosses but is not given any more clues as to the direction of the bias.

Before the first toss M reasons thus: 'The coin is definitely biased, but I don't know how much or in what direction. Thus for me any of the results $0, 1, \ldots, n$ heads on the first n tosses are equally probable.' This reasoning leads him to adopt *a priori*, the Laplacean distribution, and to choose all the necessary betting quotients accordingly. In particular, as de Finetti in fact points out (1937, pp. 143–5), his argument gives $p_r^{(n)} = r + 1/n + 2$, where $p_r^{(n)}$ is the probability of heads on the $n + 1$ the toss given r heads on the first n tosses. So in particular $p_1^{(1)} = 2/3$, i.e. if one head is observed M should change his betting quotient on heads from $\frac{1}{2}$ to $2/3$.

Suppose next that on the first toss the result is indeed heads. Suppose further that, while the first toss is taking place, M has continued to think about the problem. He now reasons thus: 'The coin did not after all look *very* bent. It is much more likely to have a small rather than a large bias. Thus I was wrong to consider each of the results r heads on n tosses as equally probable. I should have given a greater probability to results for which $r/n \doteq \frac{1}{2}$.' As a result of these new thoughts, M chooses for $p_1^{(1)}$ after the first toss, not $2/3$, but a value between $2/3$ and $\frac{1}{2}$ and much nearer $\frac{1}{2}$. His behaviour, as I have described it, seems perfectly rational throughout. Yet he does not change his betting quotients in accordance with Bayesian conditionalization.

This example can of course be generalized. We can imagine someone who, throughout a whole series of tosses, continues to rethink the problem and improve his original analysis of the situation. As a consequence, he might rarely or never change his betting quotients according to Bayesian conditionalization.

Yet I would regard such a person as highly rational – more rational indeed than someone who remained faithful to his original conditional betting quotients.

Another possibility is this. We may *a priori* consider the event A to be very unlikely to occur, and adopt the conditional betting quotient $p(E_{n+1}|A)$ *not* because we considered this a reasonable betting quotient given that A has actually occurred but merely in order to be coherent with betting situations we thought more likely to occur. If A actually did occur our whole outlook might alter, and we would want to adopt a betting quotient on E_{n+1} different from $p(E_{n+1}|A)$.

It is interesting to examine Ramsey's attitude to the present problem, for in fact he contradicts himself on this point. In (1926, p. 79) he writes: 'The degree of belief in p given q is measured thus.' The usual definition of a conditional betting quotient is given and he continues: 'This is not the same as the degree to which he would believe p, if he believed q for certain; for knowledge of q might for psychological reasons profoundly alter his whole system of beliefs.' This seems to be, in effect, a statement of Hacking's argument. Yet in the same article (1926, p. 87) we find:

> We have therefore to explain how exactly the observation should modify my degrees of belief; obviously if p is the fact observed, my degree of belief in q after the observation should be equal to my degree of belief in q given p before.... When my degrees of belief change in this way we can say that they have been changed consistently by my observation.

However, to return to de Finetti, we may supplement Hacking's fundamental criticism by some further observations. First of all de Finetti envisages a complicated system of absolute and conditional bets which are made, or at least considered, *before* the first toss. Now this is an unrealistic assumption when applied to actual betting situations. Consider for example gamblers at Monte Carlo. They bet on a spin of the wheel. The result is observed, and they then go on to make a different set of bets on the next spin, etc. No conditional bets are ever made or even considered. Yet all accept the 'objective' odds as being the correct ones. (I neglect the fact that the proprietors in fact offer unfair odds in order to increase their profits.) It thus seems

that objective probabilities can be established quite apart from
a consideration of conditional bets and Bayesian conditionalization. Moreover in such betting situations coherence only gives
us that the sum of our probabilities on the exhaustive and
exclusive set of outcomes must be 1. It does not further determine the values that we should adopt.

This argument is even more striking in the case of probabilities
in physics. Any standard textbook of atomic physics which
discusses radioactive elements will give values for their atomic
weights, atomic numbers, etc., and *also* for the probability of
their disintegrating in a given time period. These probabilities
appear to be objective physical constants like atomic weights,
etc., and, like these last quantities, their values are determined
by a combination of theory and experiment. In determining the
values of such probabilities no bets at all are made or even
considered. Yet all competent physicists interested in the matter
agree on certain standard values – just as they agree on the
values of atomic weights, etc.

We see then that the kinds of bets envisaged by the subjective
theory are unrealistic. We can further question whether such
bets would always accurately represent an individual's degree
of belief. To see this, suppose as usual that M is betting with N.[1]
N asks M to give an *a priori* betting quotient for the 1,001st toss
of a certain coin, but not to specify further betting quotients.
M examines the coin and since it looks fair gives the betting
quotient $\frac{1}{2}$. N specifies a positive stake S for the result heads on
the 1,001st toss, i.e. if heads occurs on that toss M gains $\frac{1}{2}S$; if
tails occurs, N gains $\frac{1}{2}S$. The coin is now tossed 1,000 times and
a random sequence with frequency ratio of heads of about 1/4
is observed. Naturally if *now* asked to give a betting quotient
which represented his degree of belief in heads on the 1,001st
toss, M would choose 1/4. But now N can win money off M by
setting as stake on the second bet $-S$. For in the event of heads
occurring M's total gains on the two bets are $\frac{1}{2}S - 3/4S = -1/4S$,
and if tails occurs $-\frac{1}{2}S + 1/4S = -1/4S$. So he loses in both cases.
Indeed to avoid such a Dutch book, M has to choose
$p\,(\text{heads}) \geqslant \frac{1}{2}$.

[1] The following example was suggested to me by discussions with
Mr J. Dorling. See, in this context, his article (1972, footnote 10).

What has gone wrong here? It is clear that, in the example cited, M has to choose as betting quotient *not* something that represents his true degree of belief, but something that will prevent N winning off him, granted that certain bets have previously been made. So for M's betting quotient to represent his belief, we have to neglect the effect of bets made previously. But if we neglect previous bets, then we need not change our betting rates according to Bayesian conditionalization.

My last criticism of de Finetti's argument again concerns the assumption it involves that we do not alter our *a priori* assessment of the situation in any radical fashion but only modify it by Bayesian conditioning. In fact, however, there are cases where we may want to change our initial formulation of the problem completely, and de Finetti himself even mentions such a case. To explain this I shall have to give an account of his very ingenious result concerned with exchangeable events which now goes by the name of de Finetti's theorem.

Let $X_1, \ldots, X_n, \ldots$ be exchangeable random variables and let $Y_n = 1/n \sum_{k=1}^{n} X_k$, then de Finetti first shows (1937, p. 126) that the distribution $\Phi_n(\xi) = p(Y_n \leqslant \xi)$ tends to a limit $\Phi(\xi)$ as $n \to \infty$. He goes on to say (1937, pp. 128–9):

> Indeed, let $p_\xi(E)$ be the probability attributed to the generic event E when the events $E_1, E_2, \ldots, E_n, \ldots$ are considered independent and equally probable with probability ξ; the probability $p(E)$ of the same generic event, the E_i being exchangeable events with the limiting distribution $\Phi(\xi)$, is
>
> $$p(E) = \int_0^1 p_\xi(E) \, d\Phi(\xi).$$

Thus the system of exchangeable events with limit distribution $\Phi(\xi)$ corresponds to a certain situation which might be envisaged by an objectivist; namely one in which he supposes the events $E_1, \ldots, E_n, \ldots$ to be independent with true but unknown probability ξ where ξ is a random variable lying between 0 and 1 with distribution function $\Phi(\xi)$. Similarly de Finetti thinks (1937, p. 146, footnote 4):

> One could in the first place consider the case of classes of events which can be grouped into Markov 'chains' of order $1, 2, \ldots, m, \ldots$, in the same way in which classes of exchangeable

events can be related to classes of equiprobable and independent events.

Let us call such classes of events, Markov-exchangeable events. These, according to de Finetti, would constitute a complication and extension of his theory, but do not cause any fundamental problem. Indeed, he says (1937, p. 145):

> One cannot exclude completely *a priori* the influence of the order of events.... There would then be a number of degrees of freedom and much more complication, but nothing would be changed in the setting up and the conception of the problem..., before we restricted our demonstration to the case of exchangeable events...

But now suppose we have the situation which would be described objectively as follows. We have a series of events produced by some physical process. We first think that they are independent and that the various possible outcomes have constant probabilities. Then, after examining some results we conclude that the independence assumption is inadequate and decide instead that we are dealing with a Markov chain. This Markov chain hypothesis in fact agrees with the further results obtained. Such a situation could not, I claim, be dealt with in de Finetti's system. Corresponding to the initial independence assumption, he would make an exchangeability assumption. But this *a priori* assumption once made can never, as he himself says (see the passage from 1937, p. 146, quoted on p. 17 above), be revoked. If we make a Markov-exchangeability assumption *a priori* we are all right, but once we have made the exchangeability assumption we can never change over to Markov-exchangeability in the light of the evidence. This seems plainly unsatisfactory. On the other hand if we allow a change to Markov-exchangeability in mid-channel (so to speak), then it becomes false that all changes of probabilities are by Bayesian conditionalization and the whole basis of de Finetti's argument is overthrown. I conclude that de Finetti's argument, like Keynes' appeal to the principle of indifference, fails to give an explanation, from within the logical tradition, of the existence of measurable probabilities with agreed values.

I must now make a comment on a difference between the terminology I have adopted and the more standard one. It is

3

customary to call the Ramsey–de Finetti approach the *subjective* theory and to contrast this with Keynes' *logical* theory. However, I regard the two theories as different variants within the same *logical* tradition. This is because there seems to me to be a close continuity between the analyses of probability as successively degree of partial entailment – degree of rational belief – degree of belief subject to certain weak rationality constraints. Moreover no purely subjective theory seems possible. Degrees of belief considered as irrational psychological phenomena can take on any values. In order to obtain the calculus of probabilities we must have certain rationality constraints. It is these rationality constraints which enable us to obtain numerical values for probabilities and generally speaking to use mathematical techniques. Much the same point is made by Popper in his article (1957a, p. 357):

> The subjective theory is bound to take the logical theory as its basis since it wishes to interpret $p(a, b)$ as that degree of belief in a which may be *rationally justified* by our actual total knowledge b. This is the same problem (only expressed in subjective language) which the logical theory tries to answer; for the logical theory tries to assess the degree to which a statement a is logically backed by the statement b. [My italics]

Further, my terminology is not at all incompatible with what Ramsey and de Finetti themselves say. Ramsey begins his paper: 'In this essay the Theory of Probability is taken as a branch of logic, the logic of partial belief and inconclusive argument...', while de Finetti speaks not only of the 'subjective sources' of probability but also of its 'logical laws'. We may thus define the logical approach as one which sees probability theory as an attempt to bring rational order into our system of knowledge and beliefs; which regards it as an organon or as part of a general theory of rationality.

We must now pick up another strand in the logical tradition. In 1939, shortly after de Finetti's paper, Jeffreys published his *Theory of Probability*. This, like Keynes' work, was inspired by the lectures of the Cambridge philosopher W. E. Johnson. Probability is conceived of as degree of partial entailment, and the betting quotient innovations are not adopted. Jeffreys differs from Keynes in thinking that any two probabilities are comparable. More importantly he makes a more determined effort than Keynes to tie the theory of probability in with scientific inference, and to develop the principles of statistics from his point of view. Much the same task is undertaken by Savage in his more recent *Foundations of Statistics* (1954), but Savage does adopt the betting quotient approach. A noteworthy feature of his book is the discussion of the 'minimax' principle. Finally, in the last year the literature has been enriched by the publication of de Finetti's *Teoria delle Probabilità* (1970) – a development in two volumes of his approach to the subject. Without trying to do justice to this important book, I may mention one point which has not been discussed to date. I have stated that the Dutch book argument yields the axioms of probability. This must be qualified by saying that it gives us only finite additivity, while in the standard Kolmogorov treatment countable additivity is assumed. This could be used as an objection to the betting quotient approach, but de Finetti boldly replies (1970, pp. 733–40) that countable additivity is not only unnecessary, but even undesirable.[1]

(v) Characterization of the scientific approach
It can thus be seen that the logical tradition is still vital and full of life. Nonetheless it is the other or *scientific* approach which I shall adopt in what follows. Once again my terminology differs a little from what is customary. It is usual to refer to von Mises' theory as the *frequency* theory of probability and to take as its defining characteristic the definition of probability as limiting frequency. To my mind, however, von Mises' most important idea is that of probability theory as a branch of empirical science similar to mechanics. This I take as the defining characteristic of the scientific tradition and I have adopted it as one of the mottoes of the book. I have also quoted a passage from von Mises where he describes this conception as 'the essentially new idea which appeared about 1919'. Further, many authors who agree with von Mises' 'new idea' disagree with his limiting frequency definition. For example, Cramér (1945, p. 145) writes in a manner reminiscent of von Mises:

---

[1] For a further discussion, see my review of de Finetti's book (Gillies, 1972).

> When in some group of observable phenomena, we find evidence of a confirmed regularity, we may try to form a mathematical theory of the subject. Such a theory may be regarded as a *mathematical model* of the body of empirical facts which constitute our data....
>
> Two classical examples of this procedure are provided by Geometry and Theoretical Mechanics.

The object of his book is 'to work out a *mathematical theory* of phenomena showing statistical regularity' (p. 144). On the other hand he rejects the limiting frequency definition (p. 150):

> ...some authors try to introduce a system of axioms directly based on the properties of frequency ratios. The chief exponent of this school is von Mises, who defines the probability of an event as the *limit of the frequency* v/n of that event as n tends to infinity. The existence of this limit, in a strictly mathematical sense, is postulated as the first axiom of the theory. Though undoubtedly a definition of this type seems at first sight very attractive, it involves certain mathematical difficulties which deprive it of a good deal of its apparent simplicity. Besides, the probability definition thus proposed would involve a mixture of empirical and theoretical elements, which is usually avoided in modern axiomatic theories. It would, e.g. be comparable to defining a geometrical point as the limit of a chalk spot of infinitely decreasing dimensions, which is usually not done in modern axiomatic geometry.

It is interesting to note, however, that Russell did propose to define points in a way not very different from the one Cramér describes here, cf. *Our Knowledge of the External World*, pp. 119–20.

To this criticism von Mises replied:

> ...Cramér, in his excellent book on statistics, tries to ridicule my definition of probability by comparing it to the definition of a point as the limit of a sphere whose radius tends to zero. But what is ridiculous in the latter definition is that to define a sphere and its radius it is necessary to know in advance what a point is. The concept of probability is not needed to be able to speak of an infinite sequence of arbitrary attributes and of frequency limits in such a sequence. (1952, p. 18)

and again (1963, p. 45), 'The "mixture of empirical and theoretical elements" is, in our opinion, unavoidable in a mathematical science.' He also counter-attacked in a later edition of *Probability, Statistics and Truth* by remarking (1928, p. 81): 'Cramér omits giving a clear definition of probability and in no way explains or derives in a logical manner the elementary operations of probability calculus.'

This altercation indicates one of the fundamental problems of the subject – that of elucidating precisely the relations between probability and frequency. Everyone agrees that there is a close relation between the two but there are many opinions as to how it is established. When this matter is discussed in Part II, I shall distinguish three different accounts of the matter, all held by authors within the scientific tradition. My own solution will be slightly different from any of these.

(vi) The two-concept view
So far I have emphasized the great differences between the logical and scientific approaches to probability theory. But there have of course been attempts to reconcile these two traditions. The most obvious way of doing so is to suggest that there are really two concepts of probability which have somehow got confused in ordinary language. The logical tradition deals with one of these concepts and the scientific with the other. A suggestion of this kind was made by Ramsey in his 1926 paper:

> Probability is of fundamental importance not only in logic but also in statistical and physical science, and we cannot be sure beforehand that the most useful interpretation of it in logic will be appropriate in physics also. Indeed the general difference of opinion between statisticians who for the most part adopt the frequency theory of probability and logicians who for the most part reject it renders it likely that the two schools are really discussing different things, and that the word 'probability' is used by logicians in one sense and by statisticians in another. The conclusions we shall come to as to the meaning of probability in logic must not, therefore, be taken as prejudging its meaning in physics. (p. 57.)

As far as I can judge from his fragmentary post-1926 writings, however, Ramsey did not develop this suggestion, but rather adopted a unified 'degree of rational belief' analysis of probability. Thus in his note on chance (written about 1928; see *The Foundations of Mathematics*, pp. 206–11), he denies that there are objective probabilities except in the sense that (p. 207) 'everyone agrees about them, as opposed e.g. to odds on horses'. What then are apparently objective probabilities or 'chances' as Ramsey calls them? He answers thus (p. 206): 'Chances are degrees of belief within a certain system of beliefs and degrees of belief; not those of any actual person, but in a simplified system to which those of actual people, especially the speaker, in part approximate.' And again, on p. 209: 'Probability in Physics means chance as here explained, with possibly some added complexity....'

A more recent and genuine advocate of the two-concept view is Carnap in his *Logical Foundations of Probability*. He calls the two concepts probability$_1$ and probability$_2$. Probability$_1$ is the concept which is used to appraise or assess scientific theories or predictions relative to some evidence. It is this concept we use when we say that it is probable given our past experience that the sun will rise tomorrow; or that recent experimental results render the Big Bang theory more probable than the Steady State theory. This is the sense of probability in which we speak of the probability of a hypothesis h given some evidence $e(p(h,e))$. 'Corroboration', 'confirmation', and 'support' are often used as synonyms here. Thus we might speak instead of the degree of confirmation of h given $e(C(h,e))$. Probability$_2$, on the other hand, is the concept of probability which appears *within* scientific theories and when we deal with games of chance. Thus: 'prob (heads) = $\frac{1}{2}$' or 'there is a low probability of a molecule of the gas having a velocity in the range v, $v + u$' are examples of this concept of probability.

I will reserve the term 'probability' for what Carnap calls 'probability$_2$' and sometimes speak of 'statistical' or 'physical' probability in order to emphasize this. Carnap's 'probability$_1$,' I will speak of indifferently as 'degree of confirmation' or 'degree of support' but I will never in future use the word probability in this context. The reason for this terminology lies in some ideas of Popper's.

The trouble is that the word 'probability' carries implications with which we may not wish to agree. In particular there is a well-known mathematical calculus of probabilities. If we speak of 'probability' instead of 'corroboration' we are liable naturally, perhaps almost unconsciously, to assume that degree of corroboration $(C(h,e))$ satisfies this mathematical calculus. On the other hand the calculus of probabilities was developed in connection with the study of games of chance. It is thus designed for 'statistical' probabilities and it is by no means obvious that degrees of corroboration will obey the same formal rules. The dangers involved here are well-illustrated by the following passage from Poincaré (1902, p. 183):

> ...in the two preceding chapters I have several times used the words 'probability' and 'chance'. 'Predicted facts' as I said above, 'can only be probable'. However solidly founded a prediction may appear to be, we are never absolutely certain that experiment will not prove it false; but the probability is often so great that practically it may be accepted. And a little farther on I added: – 'See what a part the belief in simplicity plays in our generalizations. We have verified a simple law in a large number of cases, and we refuse to admit that this so-often-repeated coincidence is a mere effect of chance.' Thus, in a multitude of circumstances the physicist is often in the same position as the gambler who reckons up his chances. Every time he reasons by induction, he more or less consciously requires the calculus of probabilities and that is why I am obliged...to interrupt our discussion of method in the physical sciences in order to examine a little closer what this calculus is worth, and what dependence we may place upon it.

Poincaré starts by considering predictions which are well supported by evidence. He goes on to well-founded generalizations. Thence he passes to 'the gambler who reckons up his chances' and so to the calculus of probabilities. But is the calculus developed for the gambler necessarily the appropriate tool for the physicist considering the degrees of corroboration of various hypotheses? Poincaré assumed almost unconsciously that it is, and he was followed in this by Keynes and Carnap. This assumption was, however, vigorously challenged by Popper in *Logic of Scientific Discovery* (1934).

Popper distinguishes not *two* but *three* concepts. First of all there is statistical probability and a formal calculus designed in the first place to deal with statistical probabilities. This calculus admits of another interpretation $p(h,e)$ standing for the *logical probability* of h given e or the degree to which e partially entails h. But *logical probability is not the same as degree of corroboration* $(C \neq p)$. Degree of corroboration may however be defined in terms of logical probability by means of the rather complicated formula (1934, p. 400; from Appendix IX added in 1959 edition):

$$C(h, e) = \left(\frac{p(e, h) - p(e)}{p(e, h) + p(e)}\right) (1 + p(h) p(h, e))$$

where h is consistent and $p(e) \neq 0$.

Popper does not like introducing subjective concepts like 'belief' into epistemological discussions, but I think it fair to say that he would identify 'degree of rational belief' with degree of corroboration rather than with logical probability.[1] Popper thus distinguishes 'degree of rational belief' and 'degree of partial entailment' – two notions which were identified by Keynes (cf. also Lakatos, 1968, especially section 2.2 where these matters are discussed in more detail). In what follows we will not assume the truth of these ideas of Popper's about confirmation. Nonetheless it seems important not to use probability as a synonym for confirmation at first. We thereby protect ourselves from confusion if the two notions are really distinct while leaving open the possibility that the two concepts may turn out to be much more similar than Popper suggests.

We can on the logical approach adopt a single-concept view of probability as Keynes and de Finetti do, but this is, I think, impossible on the scientific approach. Even von Mises, as we have seen, admitted that there were uses of probability not covered by his theory. He tended to dismiss these as examples of a crude pre-scientific notion. On the other hand these uses must include those we have described as 'degree of corroboration'. Now corroboration is not something that can be treated so lightly. It is surely essential to scientific rationality that we

---

[1] Compare his remark in 1959, p. 26: 'I think it likely, however, that "degree of confirmation or corroboration" – the latter term is preferable – will turn out to furnish under certain circumstances, an adequate measure of the rationality of a belief.'

adopt hypotheses which are well-confirmed. Thus the study of the circumstances under which we can validly say that a hypothesis is well-confirmed – the development in effect of a 'confirmation theory' – is a most important task for the scientific philosopher.

The preceding remarks contain the hint of an objection to the scientific approach. On this approach we have to distinguish sharply between two concepts: probability and confirmation; yet these concepts are covered by the same word 'probability' in ordinary language. Is it really plausible to suggest that we customarily confuse two quite distinct notions, and that, without any feeling of discomfort, we use the same word for two supposedly so different ideas? I do not take such a serious view of this point as perhaps an ordinary language philosopher would. On the other hand it is, I think, an important task for the upholder of the scientific approach to elucidate the relations between statistical probability and degree of confirmation. Nonetheless I will not discuss this matter further in what follows. The reason is that such an elucidation would require a detailed development of confirmation theory, and space does not permit me to discuss questions of corroboration here. I offer the same excuse for another gap in my account. It is the duty of an advocate of one point of view to criticize the opposite position. Thus I should offer a series of detailed objections to the logical theory, but again this would involve me in problems about support and confirmation, and so will not be attempted. I will therefore confine myself to the task of developing a positive account of the theory of statistical or physical probability, which is within the scientific tradition.

(vii) General outline of the book
Thus I am adopting the same approach as von Mises. I have accepted his fundamental 'new idea'. On the other hand it is obvious that I cannot accept all the details of his account. If I did, there would be no point in pursuing the matter further! However, many features of von Mises' theory *do* seem to me inadequate. I will proceed by criticizing these weak points and suggesting how they might be corrected. From this historico-critical procedure my own theory of probability will emerge. I will now describe my main disagreement with von Mises.

I have in fact two basic criticisms – one philosophical and one mathematical. These will be dealt with in the first two parts of the book. The philosophical disagreement arises like this. Von Mises wished to develop the thesis that probability was a special science, and naturally in doing so he used certain ideas about the special sciences in general. These ideas were acquired from Mach, whom von Mises greatly admired. After giving his views on the need for defining probability, von Mises adds (1928, p. 225, footnote 7): 'The best information concerning... the general problem of the formation of concepts in exact science can be found in E. Mach.... The point of view represented in this book corresponds essentially to Mach's ideas.' Von Mises later gave a glowing account of Mach's philosophy in his article 'Ernst Mach und die empiristische Wissenschaftsauffassung' (1938), and writes in a summary of his book on positivism (1940, p. 524), 'The author is a devoted disciple of *Mach....*' These tributes are indeed appropriate. As we shall see in Part I, Chapter 1, von Mises in his development of probability theory follows exactly the pattern of Mach's development of mechanics. Oddly enough this approach works better for probability theory than it did for mechanics.

Since 1919, however, the Machian or positivist outlook on the philosophy of science has been subjected to a most profound criticism in Popper's *Logic of Scientific Discovery* (1934). As we shall see the Popperian approach to philosophy of science suggests a different development of probability theory from that of von Mises. Interestingly enough Mach was also criticized by Lenin in his 1908 volume, *Materialism and Empirio-Criticism. Critical Notes on a Reactionary Philosophy.* Lenin's position is surprisingly similar to Popper's on quite a number of points. What Lenin calls 'materialism', I will refer to (in the usual terminology of Anglo-Saxon philosophy) as the 'realist view of theories'. This 'realism' or 'materialism' will be defended against Machian 'instrumentalism'.

In Part I, Chapter 1, I explain in detail the relationship between von Mises and Mach, and develop some preliminary criticisms of Mach's theory of conceptual innovation. I next try to develop an alternative theory of conceptual innovation by examining in Chapter 2 how the Newtonian concepts of 'force' and 'mass' came to be introduced. In Chapter 3 I give a general statement of my view and illustrate it by a further example drawn from thermodynamics. A couple of further topics in the philosophy of science are then discussed. The aim of the rest of the book will be to develop probability theory in accordance with the general account of the special sciences given in Part I.

In Part II I consider a more mathematical objection. In 1933 Kolmogorov, in his *Foundations of Probability*, gave a development of the mathematical theory in measure-theoretic terms. In his now generally accepted treatment, probability is introduced as an *undefined* or *primitive* notion. This modern approach makes von Mises' definition of probability and the mathematical developments associated with it look clumsy and out-dated. Here we see the motivation of Cramér's criticisms quoted earlier.

On the other hand neither Kolmogorov nor, following him, Cramér explain in an entirely satisfactory manner how the formal theory of probability is to be connected with experience. Von Mises could, and did, claim that his definition is still necessary to link the formal theory to the experimental world. On this question I partly side with von Mises. It does seem to me that a fuller account is necessary of the relations between the probability calculus and certain experimental results than is provided by Kolmogorov. As against von Mises, however, I feel that this account should be developed in a fashion which is *consistent* with Kolmogorov's modern mathematical approach.

In Part II such an account is given in detail, using the general philosophical ideas of Part I. The first step (in Chapter 4) is to relate the general questions to the particular problems of probability theory. I then give a detailed comparison of the axiom systems of Kolmogorov and von Mises. This provides me with certain important material for what follows. I now proceed by introducing the notion of a *probability system*. This is an ordered quadruplet $(\mathfrak{S}_s, \Omega, \mathfrak{F}, p)$ where $(\Omega, \mathfrak{F}, p)$ is an ordinary probability space in the sense of Kolmogorov and $\mathfrak{S}_s$ is a set of repeatable conditions whose possible outcomes are the points of Ω. In Chapter 5 repeatability is discussed and a certain axiom – what we call the 'Axiom of Independent Repetitions' – is proposed for probability systems. This axiom together with certain methodological suppositions enables us to establish the sought-for connection between probability and frequency.

The remainder of Part II is devoted to discussing a number of general questions in probability theory using the notion of a probability system. In the later sections of Chapter 5 I deal with the significance of the law of large numbers and in Chapter 6 with the definition of randomness and its role in the general theory. Finally in Chapter 7 I discuss the question of whether we can introduce, from the present point of view, probabilities for single or unrepeated events.

One of the main differences between my view of probability theory and that of von Mises can be put like this. Von Mises wished to tie in the formal calculus of probability with experience by means of a definition of probability in terms of frequency. I want to achieve the same end by an application of Popper's notion of falsifiability. However, such an application involves a number of problems and these are discussed in detail in Part III.

The main difficulty arises from the fact, stressed by Popper himself in the second of my mottoes, that probability statements are strictly speaking unfalsifiable by evidence. Nonetheless we may adopt a methodological rule, or falsifying rule for probability statements, to the effect that probability statements should in practice be regarded as falsified under such and such circumstances. In Part III, Chapter 9, I attempt to formulate such a rule, and then evaluate it in general terms in Chapter 10. The rule proposed agrees fairly well with the standard statistical tests, but at the same time conflicts with the general Neyman–Pearson theory of hypothesis testing. As the Neyman–Pearson theory is still generally accepted this bodes ill for my proposed rule, but in Chapter 11 I reply to this objection by criticizing the Neyman–Pearson theory and claiming that it, rather than my proposed rule, should be given up.

To sum up then, Von Mises tried to develop probability theory as one of the special sciences using as background the philosophy of Mach and the mathematics of such then current writers as Markov. I am attempting the same task but with the philosophy of Popper and the mathematics of Kolmogorov.

PART ONE

The Special Sciences in General

CHAPTER ONE

Von Mises' Philosophy of Science:
Its Machian Origins

My earlier description of von Mises' theory of probability suggests that he would have adopted something like the following account of special sciences in general. Suppose we wish to set up an exact or mathematical science. Our first task must be to delimit the subject matter of the putative science. Thus in geometry we deal with the spatial relations of bodies, in mechanics with their motion and states of equilibrium, in probability theory with repeated 'random' events. Because we are empiricists the next step must be to discover certain laws which are obeyed by the events or bodies with which we are dealing. Thus, for example, having devised methods of measurement we might discover that the angles of a triangle are 180°, similarly we might obtain Galileo's law that freely falling bodies have constant acceleration, and finally for probability theory we have of course the Law of Stability of Statistical Frequencies. We now cast these laws into the form of a few simple mathematical propositions. Two important points must be noted here. First, the mathematical concepts used must be precisely defined in terms of observables (the definitional thesis). Unless we do this the mathematical calculus would be unconnected with the empirical world. The concepts we use may be called by the same names as certain pre-scientific concepts of ordinary language (square, force, probability). In the theory, however, they do not have their everyday sense but a precise sense given by the definition. A second point is that the mathematical propositions do not represent reality exactly but are an abstraction or simplification of it.

Having set up our axioms we can obtain mathematical consequences of them by logical deduction. These consequences can at every stage be compared with experience using the linking definitions. Thus we check continually the empirical nature of our theory. Examples of such deductions might, in the three paradigm cases, be Pythagoras' theorem, the path of a projectile (neglecting air resistance), and the distribution of the sum of a large number of independent errors. Conversely, given any practical problem in geometry, mechanics, or probability theory we can formulate it within our calculus and use the mathematical machinery thereby put at our disposal in an attempt to solve it.

Anyone familiar with Mach's philosophy of science will recognize in the above statement the main outlines of his views. Mach's first publication in 1868 ('Über die Definition der Masse') was an attempt to introduce the concepts of Newtonian mechanics in accordance with the 'definitional thesis'. All definitions for Mach must ultimately be in terms of observables occurring in our immediate sensations. My account of von Mises' philosophy of science does omit one idea of Mach's: namely his famous claim that the purpose of science is 'economy of thought'. It is easy to see how this fits in.

We could ask: granted that the initial laws contain all the empirical truth of the theory, what is the point of introducing the mathematical calculus which anyway is an abstraction from reality? Mach's answer is that although the mathematical laws are further from reality, they summarize a great number of direct empirical facts in a very simple and concise way. They thus enable us to bring these facts easily before our minds and to apply them to the problem in hand. The purpose of the mathematical formulations is, in short, to facilitate thinking. I will now give an account of Mach's development of (classical) mechanics. This will make more explicit the relations between von Mises and Mach. It will also provide us with material for a later criticism of the Machian (and hence Misian) standpoint.

(i) Mach's development of mechanics
Mach, while professing boundless admiration for Newton, does not feel that the master's exposition of the foundations of mechanics is entirely satisfactory. He begins his criticism by quoting Newton's definition of mass (1883, p. 298):

> Definition I. The quantity of any matter is the measure of it by its density and volume conjointly.... This quantity is what

I shall understand by the term *mass* or *body* in the discussions to follow.

Mach comments (1883, p. 300):

> Definition I is a...pseudo-definition. The concept of mass is not made clearer by describing mass as the product of the volume into the density as density itself denotes simply the mass of unit of volume.

This criticism seems entirely valid. Dugas (1958, p. 342) attempts to defend Newton on the grounds that Newton was attempting to define mass in more familiar terms and granted this objective he could hardly have done better. However it is dubious whether 'density' is more familiar than 'mass', and at all events the definition does not serve the Machian function of linking the theoretical concept of mass to observables.

Mach goes on to quote Newton's definitions of force (which we will omit) and the three famous laws which for completeness we will quote:

> Law 1. Every body perseveres in its state of rest, or of uniform motion in a right line, unless it is compelled to change that state by forces impressed thereon.
> Law 2. The alteration of motion is ever proportional to the motive force impressed; and is made in the direction of the right line in which that force is impressed.
> Law 3. To every Action there is always an equal Reaction: or the mutual actions of two bodies upon each other are always equal, and directed to contrary parts.

Mach has this to say (1883, p. 302):

> We readily perceive that Laws 1 and 2 are contained in the definitions of force that precede.... The third law apparently contains something new. But...it is unintelligible without the correct idea of mass....

These points can be expanded as follows. The second law (in modern notation $\mathbf{P} = m\mathbf{f}$), Mach regards as a definition of force but one which presupposes a definition of mass. The first law is merely a special case of the second because if $\mathbf{P} = 0$, $\mathbf{f} = 0$ and only motion in a straight line with constant velocity is possible. Thus for Mach the first two laws are definitions. The empirical

4

content of the theory is contained in the innocuous-looking third law. Mach's idea is to use the experimental facts which lie behind this law to form a definition of mass. The theory of mechanics will then follow without circularity.

These 'facts' can be stated thus. Suppose we have two bodies which interact with each other. The form of interaction can be any of a large number of different kinds. The bodies may collide, they may be connected by a spring, there may be electrical or magnetic interactions between them, or finally they may be heavenly bodies attracting each other by gravity. In all these cases we observe the following law (1883, p. 303):

> *a. Experimental Proposition.* Bodies set opposite each other induce in each other, under certain circumstances to be specified by experimental physics, contrary *accelerations* in the direction of their line of junction.

This is the *Urphänomen* of mechanics, and corresponds to the Law of Stability of Statistical Frequencies in von Mises' development of probability theory. Just as that law led to the definition of probability, so this law leads to the following definition of mass:

> *b. Definition.* The mass-ratio of any two bodies is the negative inverse ratio of the mutually induced accelerations of those bodies.

An early objection to this account was that the preliminary observations on bodies necessary to establish experimental proposition *a* could only be made on an astronomical scale. How then could mechanics be terrestrial as well as celestial? Mach is much concerned to refute this (1883, p. 267):

> H. Streintz's objection...that a comparison of masses satisfying my definition can be effected only by astronomical means, I am unable to admit.... Masses produce in each other accelerations in impact, as well as when subject to electric and magnetic forces, and when connected by a string in Atwood's machine.

There are, however, further points which need to be established before the definition of mass can be regarded as adequate.

Let us suppose we have three bodies A, B and C. We measure the mass-ratio of $A:B$, $B:C$ and $A:C$ using the Machian method.

Let us suppose these come to m_{AB}, m_{BC} and m_{AC}. We obviously require that $m_{AC} = m_{AB} . m_{BC}$. However we have not postulated any experimental law that will ensure that this is the case; that will ensure, in effect, that masses are comparable. Suppose now that such a law was not true. Then by suitably arranging three masses we could obtain a perpetual motion machine of a kind which we know does not exist. This point together with a further counter to Streintz's objection is contained in Mach's experimental proposition c (1883, p. 303):

> The mass-ratios of bodies are independent of the character of the physical states (of the bodies) that conditions the mutual accelerations produced, be those states electrical, magnetic, or what not; and they remain, moreover, the same whether they are mediately or immediately arrived at.

This second experimental proposition is really little more than an expansion and completion of the first. Mach now gives a third experimental proposition which really is something new. We could compare it to von Mises' Law of Excluded Gambling Systems. It states in effect that force is a vector. Since force is later to be defined in terms of acceleration it takes the form:

> d. *Experimental Proposition*. The accelerations which any number of bodies $A, B, C, \ldots$ induce in a body K, are independent of each other. (The principle of the parallelogram of forces follows immediately from this.)

Mach now concludes his account by defining:

> e. *Definition*. Moving force is the product of the mass-value of a body into the acceleration induced in that body.

It seems to me evident that von Mises modelled his account of probability on this account of mechanics. Oddly enough, however, the approach seems to me *more* successful in the case of probability theory. The reason is this. The two experimental propositions on which von Mises founded probability theory (the Law of Stability of Statistical Frequencies and the Law of Excluded Gambling Systems) were such that they could have been and indeed had been checked in practice quite independently of the more advanced notions of probability theory. This is not the case with Mach's experimental propositions.

Consider the crucial experimental proposition *a*. In order to check this one would need to observe a large number of bodies interacting in various ways and study the accelerations they produced in each other. Had any such observations been made before Newton or had they even been made long after the acceptance of Newtonian mechanics? Certainly not. Admittedly, using various machines and devices, we could perhaps carry out these observations, but the very machines we use would have been designed in accordance with Newtonian mechanics whose foundations we are supposed to be checking! Plausible though Mach's account seems at first sight it is not clear on reflection that the experimental basis is really established without circularity.

(ii) Operationalism

Mach's idea of defining the theoretical concepts of science in terms of observables is closely connected with the view that the physical meaning of concepts is given by their operational definition in terms of the methods used to measure them. The philosophy of 'operationalism' is associated with the name of Bridgman (*Logic of Modern Physics*, 1927). We see in his book too the influence of Mach. In fact Bridgman begins by criticizing the Newtonian notion of absolute time on the grounds that the Newtonian formulation does not enable us to measure absolute time. This criticism is almost the same as Mach's (1883, p. 271, following). It is true that Bridgman quotes Einstein on this point, but, to some extent, this puts the cart before the horse as Einstein had himself been greatly influenced by Mach in his rejection of the absolutes.

After these preliminaries Bridgman comes to his main thesis which he expands in terms of the concept of length (1927, p. 5):

> To find the length of an object, we have to perform certain physical operations. The concept of length is therefore fixed when the operations by which length is measured are fixed: that is, the concept of length involves as much as and nothing more than the set of operations by which length is determined. In general, we mean by any concept nothing more than a set of operations; *the concept is synonymous with the corresponding set of operations.*

Unfortunately Bridgman is not such a systematic thinker as Mach and he develops his thesis in a confused way. This is shown most clearly in his treatment of force and mass (1927, pp. 102–8). His first suggestion is that we should define force operationally in terms of a spring balance or more generally in terms of the deformation of an elastic body. Our next development of the force concept involves considering (p. 102): 'an isolated laboratory far out in empty space, where there is no gravitational field'. In this isolated laboratory we first encounter the concept of mass. It is entangled with the force concept, but may later be disentangled. The details of this disentanglement are (p. 102): 'very instructive as typical of all methods in physics, but need not be elaborated here'. Compared with Mach's lucid account this is sheer muddle.

On the other hand, being an experimental physicist Bridgman is more concerned with the ways in which measurements are made in practice. This leads him to make a number of points which Mach did not consider and which, oddly enough, tell against the operationalist thesis. A first point is that when we extend a physical concept we have to introduce a new operational definition. Mach's definition of mass or von Mises' of probability would seem to apply to *any* masses or probabilities, but consider now the case of length. We might begin by defining length in terms of rigid metre sticks. However (1927, p. 11):

> If we want to be able to measure the length of bodies moving with higher velocities such as we find existing in nature (stars or cathode particles), we must adopt another definition and other operations for measuring length. . . .

Of course our different operational definitions must agree where they overlap, but there is another complication. Let us take the first simple extension of the concept of length. Suppose we wish to measure large terrestrial distances of the order of several kilometres, say. We have to supplement our use of metre sticks with theodolites. Now to use these instruments we have to make certain *theoretical* assumptions. For example we must assume that light rays move in straight lines and that space is Euclidean. However, as Bridgman says (1927, p. 15):

> But if the geometry of light beams is Euclidean then not only must the angles of a triangle add to two right angles, but there

are definite relations between the lengths of the sides and the angles, and to check these relations the sides should be measured by the old procedure with a meter stick. Such a check on a large scale has never been attempted and is not feasible.

But if such a check is not even feasible are we justified in making these assumptions which lie behind our operational definition? Finally, even our simple-minded definition in terms of rigid metre rods has to be subjected to a great many corrections before it can be regarded as adequate (1927, p. 10):

> We must...be sure that the temperature of the rod is the standard temperature at which its length is defined, or else we must make a correction for it: or we must correct for the gravitational distortion of the rod if we measure a vertical length; or we must be sure that the rod is not a magnet or is not subject to electrical forces.

But how are we to introduce these corrections? The case becomes worse if we remember that the concepts involved in the corrections must *themselves* be operationally defined. Does this not lead to a vicious circle? Popper for one thinks it does (1962, p. 62):

> Against this view [operationalism], it can be shown the *measurements presuppose theories*. There is no measurement without a theory and no operation which can be satisfactorily described in non-theoretical terms. The attempts to do so are always circular; for example, the description of the measurement of length needs a (rudimentary) theory of heat and temperature-measurement; but these in turn involve measurements of length.

(iii) Some objections to operationalism

Let us now examine the consequences of all this for the Machian point of view. As a matter of fact both Mach and Bridgman did at least partially realize that measurements presuppose theories. I have already mentioned Bridgman as asserting that theodolite readings presuppose theories about light rays and the geometry of space. Mach admittedly would have baulked at the use of the word theory. He says (1883, p. 271): 'All uneasiness will vanish when once we have made clear to ourselves that in the concept of mass no theory of any kind whatever is contained, but simply a fact of experience.' On the other hand he does recognize that an operational definition of mass must be based on certain laws (his experimental propositions *a* and *c*). These 'facts' are really universal laws and can justly be referred to as theories.

We are now in a position to define the Machian operationalist position more precisely, and in future we will use the word 'operationalism' only in this restricted sense. What it amounts to is this: every new concept introduced into physics must be given an operational definition in terms of experimental procedures and concepts already defined. The empirical laws which lie behind these definitions must be established by observations before introducing the new concept. Bridgman's various points about experimental method raise two objections against 'operationalism' in this sense.

First of all, one single operational definition does not suffice for most concepts. As the use of the concept is extended to new fields, it must be given new operational definitions. It is very difficult to see how the laws on which the operational definition is based can be verified without considering the new concept itself. Imagine, for example, verifying that space is Euclidean before introducing the concept of length! I shall call this: the objection from conceptual extension.

Secondly, there is an objection concerned with the correction and improvement of methods of measurement. Suppose we introduced a naive definition of length in terms of rigid metre rods and employed it to measure lengths up to say half a kilometre. Then the theodolite method is discovered. At once it is employed for lengths of more than 50 metres. Now normally we would say that a *more accurate* method of measuring lengths more than 50 metres had been discovered. On the operationalist view, however, this manner of speaking is inadmissible. We have *defined* length by the rigid metre rod procedure and the most we can say of another method of measurement is that it gives results in approximate agreement with the defining procedure for length. It makes no sense to say that the results given by the alternative method are nearer to the true value of the length than those given by the defining method. That would be like first *defining* a metre as the distance between these two marks on this rod and then saying that more accurate measurement had revealed that the distance wasn't a metre.

The situation is the same when we come to consider the 'corrections' for temperature, gravitational and electrical distortions, etc. mentioned earlier. Suppose again we had defined length in terms of a measuring procedure using iron rods but without taking temperature corrections into account. One day bright sunshine falls through the windows of the laboratory heating both the measuring rod and the wooden block being measured. It is observed that relative to the rod the wooden object has changed its length from the day before (in fact contracted). However, an intelligent experimenter then suggests that in fact the measuring rod has expanded more than the wooden block. He cools down the rod to normal room temperature and produces a more correct value of the new length of the block. Indeed he now shows that it has expanded rather than contracted. But how is this admissible on the operationalist point of view? Length has been defined by the initial set of procedures and according to this definition the block must have contracted rather than expanded.

The only line the operationalist can take on this is to say that we have decided to adopt a new definition of length. Our naive rigid-metal-bar definition is replaced for distances over 50 metres by a theodolite definition, while in certain other circumstances a temperature correction is introduced. But the operationalist now has to give an account of how new definitions are evolved and why we choose to adopt one definition rather than another. Further, in view of Popper's point he has to show that the new definitions do not involve circularity since many of the correcting terms must themselves be given operational definitions.

These two objections indicate the grave difficulties which stand in the way of any systematic operationalist account of the introduction and development of the concepts of physics. Mach's definition of mass, for example, gives only the barest beginnings of such an account. These difficulties are, I believe, insuperable.

Having criticized operationalism, it is now worth pointing out that it is an attempt to solve a serious and difficult problem in the philosophy of science. This difficulty can be described using the notion of 'empirical meaning' or 'empirical significance'. Let us say that a concept has 'empirical meaning (or significance)' if we can assign numerical values to particular instances of it – if we can, in effect, measure it under certain circumstances. Thus the concepts of force and mass have empirical meaning because we can, at least in some cases, measure the masses of bodies and the forces acting on them. But now we can ask: how can new concepts acquire empirical significance? If they are not defined in terms of observables or by means of the methods used to measure them, how *do* they acquire meaning? This I shall call 'the problem of conceptual innovation'.

My method of tackling this problem will be to give a detailed historical analysis of the introduction of the Newtonian concepts of force and mass. In the next chapter I will begin by outlining the background knowledge in astronomy and mechanics against which Newton developed his theory. I will then discuss how the theory was tested initially, paying particular attention to the rôle played by the new concepts of force and mass in these tests. Finally I will consider how forces and masses came to be measurable. I can then in Chapter 3 generalize from this example to give a theory of conceptual innovation in the exact sciences. It will be shown that this theory avoids the difficulties which we have noted in operationalism, and it will be explained how we can apply it to the problem of introducing the concept of probability.

CHAPTER TWO

Force and Mass

(i) The concepts of force and mass before Newton

Our aim is to study the general problem of conceptual innovation by examining Newton's introduction of the concepts of 'force' and 'mass'. It could first be asked, however: 'Is this example of conceptual innovation a genuine one? Did not some idea of "force" and "mass" exist before Newton?' Well of course *some* idea of these concepts did exist but very little, so that the example is a surprisingly good one. We can appeal to the authority of Mach on this point. He says (1883, p. 236):

'On perusing Newton's work the following things strike us at once as the chief advances beyond Galileo and Huygens:

(1) The generalization of the idea of force

(2) The introduction of the concept of mass....'

Naturally, however, an appeal to authority is not a very satisfactory method of argument so we will attempt a brief survey.

Let us first take the concept of mass as distinct from weight. In a sense, of course, Descartes drew a clear distinction between mass (or quantity of matter) and weight. He identified matter with spatial extension and thus quantity of matter was measured by volume. It naturally followed that quantity of matter was not proportional to weight. Indeed a vessel when filled with lead or gold would not contain more matter than when 'empty', i.e. filled with air (Descartes' own example). It seems to me that this concept of 'quantity of matter' is too different from Newton's to be considered as forerunner of the latter. Some authors have given Huygens the credit for being the first to distinguish 'mass' and 'weight'. In his treatise *De Vi Centrifuga* Huygens says: 'the centrifugal forces of unequal bodies moved around equal circumferences with the same speed are among themselves as the weights or solid quantities – inter se sicut mobilium gravites, seu quantitates solides' (quoted from Bell, 1947, p. 118). It has been suggested that this is the earliest hint of a distinction between mass and weight. There are also two notes in Huygens' manuscripts of 1668 and 1669 (Bell, 1947, p. 162), namely:

'(a) Gravitatem sequi quantitatem materiae cohaerentes in quolibet corpore.

(b) Le poids de chaque corps suit précisement la quantité de la matière qui entre dans sa composition.'

For my part I find the *De Vi Centrifuga* quotation unconvincing. Huygens could simply be using 'solid quantity' as a synonym for 'weight'. The manuscript quotations are more striking. However, at all events these ideas of Huygens can be ignored when taking account of the background of Newton's thought. Although the *De Vi Centrifuga* was composed around 1659, it was not published till 1703, while the Huygens' manuscripts were not published till our own time.

The third possible claimant to the concept of mass is Kepler. In the introduction to the *Astronomia Nova* (1609) he says (quoted from Koestler, 1959, p. 342):

If two stones were placed anywhere in space near to each other, and outside the reach of force of a third cognate body, then they would come together, after the manner of magnetic bodies, at an intermediate point, each approaching the other in proportion to the other's mass (moles).

This remarkable passage contains already the principle of universal gravitation, but, as Koestler rightly points out, it remained an isolated insight. Kepler later developed his celestial dynamics on other principles. Thus to a first approximation at least we can say that the concept of mass as distinct from weight is original to Newton.

The case of force is not so clear cut. We must admit that in the study of statics and equilibrium a notion of force had evolved, but it was little more than a slight generalization of the idea of weight. As Mach says (1883, p. 57): 'Previous to Newton a force was almost universally conceived as the pull or pressure of a heavy body. The mechanical researches of this period dealt almost exclusively with heavy bodies.' The two main statical laws that had been discovered at that date, namely the law of the lever and the law of the inclined plane, can in fact be stated using only the notion of weight. On the other hand the idea of weight had been generalized to give the notion of a tension in a string. This appears, for example, in the second day of Galileo's *Two New Sciences* (1638, p. 122).

In 1672, at the end of his *Horologium Oscillatorium*, Huygens published thirteen propositions without proof on centrifugal force. He considered a centrifugal force as a real force which balances the tension in a string. We can therefore take Huygens' concept of centrifugal force as a generalization of the previous notion of statical force. However, once again I am inclined to exclude Huygens' work from an enumeration of the background to Newton's thought. Admittedly the thirteen propositions were published in 1672 long before the *Principia*. However, as Herivel (1965) has shown from a consideration of early manuscripts, Newton evolved his own ideas of centrifugal force in the period 1666–9 and independently of Huygens.

In dynamics too there had been some notion of force. Galileo had worked with a concept of impeto – no doubt derived from the medieval thinkers. But this notion – in so far as it was quantitative – corresponds more closely to the modern notion of momentum than to the Newtonian idea of force. Again, Kepler has a theory of bands of force or influence emanating from the Sun and carrying the planets round like the spokes of a wheel. However these ideas of Kepler's and indeed of Galileo's were never put on a quantitative basis and were not needed in the statement of these authors' main results. I conclude that Newton's *quantitative* notion of *dynamical* force was indeed original to him. More generally we can say that we do genuinely have here a case of conceptual innovation and a careful study of it should tell us a great deal about the way in which new concepts can be developed.

(ii) How Newton introduced the concepts of force and mass

We can now state the main quantitative results which had been achieved before Newton in mechanics and astronomy. These were of course Kepler's and Galileo's laws. Despite their familiarity it might be worth repeating them briefly. Kepler's laws are three in number:

(a) Every planet moves in an ellipse with the Sun at one focus.

(b) The radius vector from the Sun to a planet sweeps out equal areas in equal times.

(c) If a is the mean distance from the Sun to a planet and T is the time of a full revolution of the planet (the length of the planetary year), then $a^3/T^2 = $ constant.

The only point worth making about these laws is that they were mixed up in Kepler's work with a great deal that was incorrect. In particular, of course, Kepler had a theory about the relation of the solar system to the five regular solids, and he considered *this* theory to be his greatest scientific achievement – much finer than the three laws. It therefore required considerable selectivity on Newton's part to obtain *just* those three laws from Kepler.

Galileo's results can be summarized very conveniently into two laws, namely:

(a) Neglecting air resistance, freely falling bodies have a constant downward acceleration g.

(b) Neglecting air resistance again, bodies which are smoothly constrained to move at angle α to the horizontal (e.g. by an inclined plane) have an acceleration $g \sin \alpha$. To these main results we may add numerous astronomical observations concerning the Moon, which, as we shall see, proved important.

In this statement I have once again rather over-simplified. Other results were known in mechanics – for instance the laws of impact. I think, however, that it is better to omit these as their inclusion would only complicate the discussion without adding any new point of importance.

We are now in a position to examine how Newton's theory was checked against experience prior to its acceptance. In doing so we must not fall into the error of supposing that we can test out Newton's laws separately. Indeed this error was committed by Newton himself because, after each law in the *Principia*, he gives the 'experimental evidence' for it. However in fact the three laws of motion *and* the law of gravity can, in the first instance, only be matched against experience *all together*. This I believe is the source of much of the misunderstanding of Newtonian mechanics. In most of the standard expositions the three laws of motion are first introduced and their consequences for problems of terrestrial mechanics are dealt with. Then in a separate section (usually the final chapter or even an appendix to the book) the law of gravity is stated and some of its astronomical consequences are mentioned. This conceals the fact that Newton's four laws form a unified cosmological theory and that they were tested out in the first instance on an astronomical scale. Essentially the theory was checked against experience by showing that all the previous results in mechanics, i.e. Kepler's laws, Galileo's laws etc., could be shown to hold to a high degree of approximation if the theory were true. We must now examine how the concepts of 'force' and 'mass' were used in this deduction.

Newton's theory can be summarized in the familiar vector equations:

$$\mathbf{P} = m\mathbf{f} \qquad \text{(which contains the three laws of motion)}$$

and $\mathbf{F} = \dfrac{\gamma m_1 m_2}{r^2} \hat{\mathbf{r}}$ (the law of gravity).

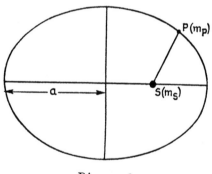

Diagram 1

Let us apply these equations to a planet P, mass m_P, moving round the sun S, mass m_S. We first neglect the gravitational interactions holding between the planets themselves. The problem is then reduced to a two-body problem, and we obtain that P moves on an ellipse of major semi-axis, a say. If the period of its orbit is T, then

$$a^3/T^2 = \gamma(m_S + m_P)/4\pi^2. \qquad (1)[1]$$

We now assume that the mass of the Sun is very much greater than that of the planet ($m_S \gg m_P$) and so obtain

$$a^3/T^2 \doteqdot \gamma m_S/4\pi^2 \qquad \text{(i.e. constant).}$$

[1] A proof of this result is given in Rutherford (1951), pp. 66–71.

This is an approximate version of Kepler's third law. The assumption $m_S \gg m_P$, though automatically made and easily overlooked for this reason, contains the solution to the problem we have been discussing. Do we need an operationalist definition of mass at this point? Not at all. We test out our theory involving masses by making the qualitative physical assumption that one mass is very much greater than another. Moreover this qualitative assumption is justified by a crude (or intuitive) notion of mass. If we think of mass as 'quantity of matter', then observing that the Sun is very much larger than the planets and making the reasonable postulate that the density of its matter is at least comparable to that of the matter in the planets we obtain that $m_S \gg m_P$. So we do not at first need a precisely defined notion of mass. A rather crude and intuitive notion of mass can lead to a qualitative assumption and so to a· precise test of a theory involving an exact idea of mass.

Let us now examine how approximations to Galileo's laws are obtained. Write M for the mass of the earth, R for its radius, and m for the mass of a small body at height h above the earth's surface. We now have to use a theorem of Newton's that we can replace a sphere whose mass is distributed with spherical symmetry by a mass point at the centre of the sphere for the purpose of calculating gravitational forces. This theorem incidentally was one which gave Newton a great deal of trouble. He only succeeded in proving it in 1685. It now appears as *Principia*, Book 1, Prop. 76.[1] Using it we have that the mass m is, by the law of gravity, acted on by a force directed towards the centre of the Earth and of magnitude $\gamma m M/(R+h)^2$ where γ is the universal constant of gravitation. As $R \gg h$ we can write this approximately as $\gamma m M/R^2$. Now $\gamma M/R^2$ is a constant for all bodies m ($= g$ say). Therefore a body m is acted on by a downward force approximately equal to mg. Since $\mathbf{P} = mf$ this gives a downward acceleration of approximately g in free fall. If on the other hand the body is acted on by a smooth constraint at angle α to the horizontal, the component of the force at angle α is approximately $mg \sin \alpha$ (since force is a vector). Therefore the body's acceleration is approximately $g \sin \alpha$. In this way approximations to Galileo's two laws follow.

[1] A proof using modern methods can be found in Rutherford (1951), pp. 25–30.

It will be noticed that in this derivation, too, an approximation is made – namely $R \gg h$. However, this is not an assumption involving new concepts, but a statement involving two quantities which were already quite well-known. After all, the first measurements of R date back to Eratosthenes. We see that in this case the laws are derived by *completely eliminating* the new concepts in the deduction. It might be asked: could we test out a new theory with new concepts by the following method: make a series of deductions from the theory in which the new concepts are completely eliminated and compare the results we obtain with experience? I myself do not think so. We might derive some of our results (like the approximations to Galileo's laws) in this way; but I think we need to obtain at least some of the others by qualitative assumptions involving new concepts (e.g. $m_S \gg m_P$). My reason is this: If the new concepts could always be eliminated before a comparison with experience took place, we would be inclined to regard the new concepts not as being physical quantities, but rather as mathematical coefficients introduced to make the calculations easier. A good example of such a 'mathematical coefficient' is $(-1)^{1/2}$ as it is used in the theory of electrical circuits. Here we always write the current i in the complex form $i_0 \exp(-1)^{1/2} \omega t$. This mathematical device greatly simplifies all the calculations. Yet we never give a physical meaning to the imaginary part of the expression. At the end of the calculation the real part of $i_0 \exp(-1)^{1/2} \omega t$ is taken and compared with the experimental findings. However, the concept of mass is not in this position.

(iii) The 'moon-test' of the law of gravity

The deduction of the approximate truth of Kepler's and Galileo's laws provided the main evidence for Newton's theory. However it is interesting also to consider the test of the theory made by observing the motion of the Moon. This moon-test was in fact Newton's first test of the law of gravity, and in the course of considering it we can examine Newton's own attitude to the questions we have just been discussing. The logic of the test (using modern methods) is as follows:

Let the Earth E have radius R and mass M. Let the radius of the Moon's orbit be a and its period (i.e. the lunar month) be T.

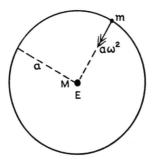

Diagram 2

If the Moon's angular velocity is ω its centripetal acceleration is $a\omega^2$ and the gravitation force on it is $\gamma Mm/a^2$. Therefore since $\mathbf{P} = m\mathbf{f}$,

$$\gamma Mm/a^2 = ma\omega^2.$$

But

$$\omega = 2\pi/T;$$

therefore

$$\gamma M/4\pi^2 = a^3/T^2. \tag{2}$$

On the surface of the Earth we have by a previous calculation

$$\gamma Mm/R^2 \doteqdot mg.$$

Therefore

$$\gamma M \doteqdot gR^2$$

substituting we obtain

$$g \doteqdot (4\pi^2/R^2)(a^3/T^2).$$

Now all the quantities in the r.h.s. of this equation are known. So we may calculate the value of g from the equation, and this can be compared with the value of g observed by means of pendula. When Newton first performed this test (c. 1666), he found a noticeable discrepancy between the two values of g. This led him to abandon his theory for a while. However the disagreement was due to a faulty value of the Earth's radius. When he tried the test again much later (between 1679 and 1684) with a corrected value of the Earth's radius, it gave agreement within experimental error and this was one of the factors which stimulated him to push his work on the *Principia* through to its completion.

It looks as if the deduction just made is similar to the deduction of Galileo's laws, i.e. the new concepts simply cancel out. However this is not in fact so. The equations are derived on the assumption that the Earth is fixed. *Eppur si muove*, and we therefore should substitute for (2) our equation (1), i.e.

$$\gamma(M + m)/4\pi^2 = a^3/T^2.$$

To regain (2) and carry through the deduction, we must again assume that m is negligible compared with M (i.e. mass(Earth) $\gg$ mass(Moon)). So the case is really the same as the deduction of the approximation to Kepler's third law.

We can now examine what Newton himself says about this. He describes the moon-test in *Principia*, Book III, Prop. 4, and after giving a deduction equivalent to our original one (but using his own mathematical methods) he continues (1687, p. 409):

> This calculus is founded on the hypothesis of the earth's standing still; for if both earth and moon move about the sun, and at the same time about their common centre of gravity, the distance of the centres of the moon and earth from one another will be $60\frac{1}{2}$ semidiameters of the earth: as may be found by a computation from Prop. LX, Book I.

Book I, Prop. 60 in effect introduces the corrected equation (1) instead of the original equation (2). We can, I think, criticize Newton's logic here. To introduce the corrections he speaks of we need to know the value of the ratio of the Moon's mass to the Earth's mass. Now this ratio can be calculated once Newton's theory is assumed by a method which I will explain in a moment. However, when Newton's theory is being given its first tests prior to its acceptance we cannot introduce the exact correction. In fact in order to get the test at all we have to introduce the qualitative assumption mass(Earth) $\gg$ mass(Moon), as has already been shown.

Similar criticisms can be raised against Newton's general method in Book III of the *Principia*. He begins by stating Kepler's laws, the motions of the Moon and of the satellites of Jupiter and Saturn as Phenomena 1–6. Then using his rules of reasoning he infers the law of gravity inductively in Book III, Props. 1–5 and 6. Finally, assuming the law of gravity he derives

Kepler's laws deductively in Props. 13–16, but this time he incorporates certain corrections. Thus he says at the beginning of Prop. 13 (1687, p. 420): 'We have discoursed above on these motions from the Phenomena. Now that we know the principles on which they depend, from these principles we deduce the motions of the heavens *a priori*.' He then goes on to mention some of the corrections which must be introduced. He claims that broadly speaking 'the actions of the planets one upon another are so very small that they may be neglected...' (p. 421). However, he goes on to say that Jupiter and Saturn noticeably affect one another in conjunction, and that the orbit of the Earth is 'sensibly disturbed by the moon' (p. 422).

Newton does not realize that these corrections actually vitiate his inductive-deductive approach. They vitiate it because, as Duhem was the first to point out,[1] the law of gravity is strictly speaking inconsistent with the phenomena from which it was supposedly induced. Moreover it is very implausible to claim that from given premises we can induce conclusions which logically contradict the premises. Newton's approach also conceals the rôle which the new concepts 'force' and 'mass' play in the derivation of approximations to Kepler's and Galileo's laws: the rôle I have tried to analyse in this section.

Duhem's point can be used to provide an additional argument against Mach's philosophy of science. According to Mach, high-level mathematical theories (such as Newton's) are merely summaries of experimental laws, and are introduced for 'economy of thought'. Thus presumably Newton's theory is a summary of Kepler's laws, Galileo's laws, the laws of impact, and perhaps other things. But this is not in fact so because Newton's theory, far from summarizing e.g. Kepler's laws, strictly speaking contradicts them. Only a certain approximation to Kepler's laws follows from Newton's theory. We could say that Newton's theory corrects Kepler's original laws; but this is unaccountable on Mach's position. After this brief digression let me complete my account of how new concepts came to be measurable.

(iv) How force and mass came to be measurable

Once a new theory involving new concepts has passed a number of preliminary tests, we can accept it provisionally and use it

[1] See Duhem, 1906, Part II, Chapter VI, Part 4, pp. 190–5.

to devise methods for measuring the values which the new concepts assume in certain particular cases. I will now illustrate this in the example of Newton's theory which we are considering. First let us consider how the mass of a planet might be measured, assuming for the moment that the Sun has unit mass. Let the planet be distant a_P from the Sun and have orbital period T_P. Then assuming $m_S \gg m_P$ we have as usual:

$$a_P^3/T_P^2 \doteq \gamma m_S/4\pi^2.$$

But now suppose the planet has a moon M of mass m_M which is distant a_M from the planet and has orbital period T_M. If we assume again that $m_P \gg m_M$, we get as before

$$a_M^3/T_M^2 \doteq \gamma m_P/4\pi^2.$$

Therefore dividing, we obtain

$$m_P/m_S \doteq (a_M/a_P)^3 \, (T_P/T_M)^2.$$

All the quantities on the r.h.s. of this equation can be determined by astronomical measurement and so we obtain a value for the ratio m_P/m_S. Calculations of this sort are given in the *Principia*, Book III, Prop. 8.

For example, Mars has a moon Deimos whose period is 30·3 hrs. We hence obtain $m_{\text{Mars}} : m_{\text{Sun}} = 3 \cdot 4 \times 10^{-7}$. Incidentally this calculation confirms our original assumption that mass (Mars) $\gg$ mass (Sun), but it is worth noting that to make it, we have to assume not only Newton's theory but also that mass (Deimos) $\ll$ mass (Mars).

As regards measuring the masses of terrestrial bodies, the case is so trivial that it is hardly worth mentioning. We identify the downward gravitational force on a body with its weight. But since the downward force is mg and g is constant we obtain mass $\propto$ weight. This gives us a method of measuring the masses of bodies. The only point worth mentioning is that the theory enables us to correct for the variation in g, which can itself be measured by means of pendula.

I will now attempt to generalize from this example of Newton's theory to obtain a general account of conceptual innovation in the exact sciences. It can then be pointed out that this account avoids the difficulties inherent in operationalism. These matters and some others will occupy us in the next chapter.

CHAPTER THREE

Conceptual Innovation in the Exact Sciences

(i) Another illustration of conceptual innovation : the concept of temperature

Let us now develop the ideas we have acquired from the case of mechanics by applying them to the problem of introducing the concept of temperature. This time no attempt will be made at a historical analysis, but I will confine myself to giving a hypothetical series of theories and tests which would have enabled a precisely measurable concept of temperature to evolve without circularity. Some details of the actual history may be found in Roller (1957). My suggestion is that we begin by proposing the following law : 'For rods or columns of a large number of different materials $\theta \propto l$, where θ is the temperature of the rod or column and l its length.' Now the interesting thing about this law is that 'θ' and 'l' both stand for *new* concepts. We are not assuming any prior notion of length, for the measurement of length, as Popper says, 'needs a (rudimentary) theory of heat' (Popper, 1962, p. 62. Quoted earlier on p. 44). But if both θ and l are new concepts, how can we test the law? The case seems altogether hopeless.

It is not, however, as hopeless as it seems. Once again we proceed by making a series of qualitative assumptions of the form : 'In such and such circumstances the temperature of this body is approximately the same as the temperature of that body', 'the lengths of these two bodies are nearly the same', etc. These assumptions enable us to obtain certain results which can be compared with experience, thereby testing out our law $\theta \propto l$. We will now analyse how this comes about.

Our first step is to select two fixed points on the temperature scale. These are of course melting ice and boiling water. It is assumed (i) that these two points represent approximately constant temperatures and (ii) that any body immersed for a sufficiently long time in the melting ice or boiling water will acquire approximately at least the *same* temperature as the melting ice resp. boiling water. Assumption (ii) plays much the same rôle as the assumption that $m_0 \gg m_1$ in the Newtonian case. We can now make a first crude test on our law. If it is assumed that the constant of proportionality in $\theta \propto l$ differs for different materials then we will expect that rods or columns which have the *same* length in melting ice (which we can call $\theta = 0$) will have different lengths in boiling water ($\theta = 100$). This can be checked. To do so we need not have a general method of measuring length but only an ability to check that two lengths are approximately the same (by putting them end to end), and to judge that one length is greater than another.

Of course we have not really checked the relation $\theta \propto l$, only that length and temperature vary together and at different rates for different materials. However our results show that certain materials, e.g. wood, show very little variation even between temperatures as different as 0 and 100. We now assume that room temperature varies very much less over a period of a few weeks than the difference between 0 and 100. So if we fix on some standard distance, we can construct a rough instrument for measuring length, viz. a ruler.

We have almost reached a position where $\theta \propto l$ can be tested but there are still difficulties. We need to use materials for which the variation is large, and we have to ensure that though the material itself is at various temperatures the length-measuring instrument, i.e. the ruler, is at room temperature (assumed approximately constant throughout the experiment). The way in which these difficulties are overcome is well-known. We choose for our materials different liquids (mercury, alcohol, etc.) encased in thin tubes of glass, closed at one end and terminating in bulbs at the other. The bulb is immersed in the melting ice, boiling water, etc. but the ruler is held against the glass tube at room temperature. It may now be objected that we can only calibrate the thermometer by assuming linearity. On the other hand, I reply, we can first test out various consequences of the law $\theta \propto l$. If these are satisfied, we assume the law and use it for our calibration. What then are these preliminary tests?

Let t stand for room temperature, assumed constant throughout the experiment. Let us consider a particular material, say a column of alcohol, and suppose that for it $\theta = kl$. Let us measure its length using the ruler at 0, t, 100 and obtain l_0, l_t, l_{100}. Then

$$100 = k(l_{100} - l_0)$$
$$t = k(l_t - l_0)$$

Therefore

$$t = 100(l_t - l_0)/(l_{100} - l_0)$$

Therefore for all substances we have (on a given day)

$$100(l_t - l_0)/(l_{100} - l_0)$$

is approximately constant. We have here a consequence which can be tested with the crude means at our disposal. Further, we can vary the experiment by taking another fixed point, say the temperature (τ) of a mixture of ice, salt and water, and checking that again, $100(l_\tau - l_0)/(l_{100} - l_0)$ is approximately constant for all materials.

We see that once again no operational definition of length or temperature is necessary. We introduce a hypothesis or theory involving these concepts, and test it out in certain ways. An instrument for measuring temperature, viz. the thermometer, is then designed on the basis of the theory. In order to make the tests we have to add to our theory certain qualitative assumptions about the new concepts. The only difference from the Newtonian case is that *there* the qualitative judgments were judgments of inequality ($m_0 \gg m_1$), whereas here they are judgments of equality viz. (a) the temperature of certain processes, i.e. melting ice and boiling water, is approximately constant, (b) the temperature of two bodies which have been immersed for a long time in boiling water is approximately the same and equal to that of the boiling water, and (c) two rods are the same length (at the temperature in question) if one can be exactly superimposed over the other.

Let us now pursue the development of the temperature concept a little further. Having obtained a method of measuring temperature (the mercury thermometer) we can now test some other laws involving temperature, for example the gas law $PV = RT$. This law holds very well for gas at very low pressures and we may therefore use it to design a very accurate (though cumbersome) instrument for measuring temperature – the so-called ideal gas thermometer. We can use this instrument in turn to test out further laws – say the thermocouple effect; and this effect can be used in its turn to provide a method for measuring small temperature differences. Now we come to an interesting point. Using our thermocouple we can test out one of our original assumptions – say that two bodies immersed in boiling water have the same temperature. We may well find that this assumption holds only approximately but not exactly. Our new methods of measurement transcend our original crude ones; but, on the other hand, the original crude assumptions and methods were necessary before the sophisticated and exact methods could be developed.

I have two analogies to illustrate this situation. The first one concerns the process of liquifying a gas. One standard method here is to use the Joule–Kelvin effect. On the other hand the Joule–Kelvin effect will only cool the gas further if it is already at a sufficiently low temperature. Let us suppose that the gas is initially above this critical temperature. It must then be cooled below it using some method less sophisticated than the Joule–Kelvin effect. Similarly, we sometimes have to use a cruder method of measurement to test out the theories on which a more sophisticated method of measurement is based. Another analogy is with the method of finding numerical values for the roots of equations by successive approximation. Usually there is an iterative process. We start with some very crude approximation and by applying a certain procedure we obtain a better value. This value is then the starting point for a new application of the procedure, etc. After several repetitions we may obtain a very accurate value but this was only possible because of our initial crude approximation.

There is another point worth raising here. We may well discover that our original crude assumption and crude law ($\theta \propto l$) do not hold exactly. Indeed we will naturally hope to correct them because in so doing we will have improved on the situation which held before. On the other hand, suppose that we show that our original laws and assumptions are not just inexact but *wildly wrong*. If this turned out to be the case we would be in an embarrassing situation. A kind of contradiction would have arisen in the notion of temperature and we would have either to abandon the concept completely or reconstruct it painfully from crude beginnings. This example shows the need for some kind of generalized principle of correspondence in scientific progress. We will return to this point in a moment, but next let us state in general terms our theory of conceptual innovation.

(ii) General theory of conceptual innovation

Let us suppose a new theory is proposed involving new concepts $C_1, \ldots, C_n$. Our problem is: how do these concepts become measurable? How do they acquire empirical significance? The answer is this. We first test the new theory by deducing from it consequences which do not involve the new concepts and comparing these consequences with experience. In some cases the deduction is strict and the new concepts are eliminated by purely logical moves without making any additional assumptions. This was the case with Galileo's laws. However, not all the consequences can be obtained in this way, otherwise the new concepts will be regarded as mathematical auxiliaries similar to $(-1)^{1/2}$ rather than as concepts with physical significance. In general, certain qualitative assumptions of approximate equality or of great inequality in particular physical situations will be made concerning the new concepts. The original theory together with these qualitative assumptions will lead to the conclusion that certain consequences hold approximately. These consequences are then matched against the results of experiments past or future. If the new theory is corroborated by these comparisons, it is accepted and methods for measuring the new concepts are devised on the basis of it. In this way the concepts acquire empirical significance. At a later stage, the original theory, or the qualitative assumptions, may be tested using more sophisticated methods of measurement and may be found to hold only approximately. The more sophisticated methods could not, however, have been developed without the previous cruder ones.

(iii) How our theory of conceptual innovation avoids the difficulties in operationalism

At the risk of being a little repetitious I will now point out that this theory avoids the difficulties in operationalism which were noted earlier. The first problem I called the problem of conceptual extension. It was observed that as a concept is extended into new fields we need new operational definitions. The laws on which these new operational definitions are based must be verified 'before introducing the concept itself'. The simple example I gave of this was extending the rigid metre rod definition of length by using a theodolite. However, the theodolite is based on Euclidean geometry whose truth must apparently be verified before introducing the concept of length.

My main disagreement here is that we regard concepts as acquiring meaning *not* through operational definitions, but through their position in a nexus of theories. An account of the logical relations of these theories and of the way we handle them in practice would give us the significance of the concept. Thus a concept can indeed be extended, *not* by acquiring new operational definitions, but rather by becoming involved in a series of new and more general theories. If we accepted the operationalist view, we could not suddenly postulate a new theory with new concepts. The new concepts would only have meaning after they had been operationally defined. An operationalist must therefore check the laws on which his definitions are to be based *before* introducing the concept. I described earlier Mach's attempt to check certain mechanical laws before introducing the concept of mass. In general, however, this programme cannot be carried through as we saw from the absurdity of checking Euclidean geometry without introducing the notion of length. Moreover from our point of view it is unnecessary. We are quite free to introduce a new undefined concept in a new theory. Our only problem then is how to test this theory and this problem can, as we have seen, be solved.

The second difficulty in operationalism was the question of how the operationalist could give an account of the correction and improvement of methods of measurement. We often, for example, speak of 'discovering a more accurate method of measuring a concept' but if the previous method was the *definition* of the concept how is any more accurate method of measuring it possible? Again we often introduce corrections for temperature, gravitational forces, etc. But how can we correct a definition?

This difficulty too disappears as soon as we recognize the

primacy of theories. Methods of measurement are only introduced on the basis of theories; and there is no reason why,
starting from a particular set of theories, we should not be able
to devise two methods of measurement – one more accurate
than the other. Again, our methods of measurement involve not
only the general theories but also certain qualitative assumptions, e.g. that temperature variations in the laboratory are
negligible. We can always replace such an assumption by a more
sophisticated one, thus 'correcting' our previous method of
measurement.

Our theory further suggests the following methodological rule
which can be regarded as an important *principle of empiricism*:
'Every new scientific theory or hypothesis must be tested before
it is accepted'. It should be noted that we are here distinguishing
between 'hypotheses and theories' and 'auxiliary assumptions'.
I have argued that in order to test a theory we need to make
certain additional assumptions, e.g. that

$$\text{mass (Sun)} \gg \text{mass (Earth)}.$$

Now these assumptions can be considered as theories, and if we
demanded that they should be tested before being accepted we
would be landed in an infinite regress and no testing would be
possible. However, if we distinguish between 'theories' and
'auxiliary assumptions', then we can apply the principle of
empiricism to the theories themselves. Of course the auxiliary
assumptions can be tested out at a later stage. For example, if
we assume Newton's theory, we can calculate from observations
of Mars' moon Deimos that mass (Mars) : mass (Sun) = $3{\cdot}4 \times 10^{-7}$.
This calculation confirms our original assumption but it is worth
noting that to make it we have to assume not only Newton's
theory but also that mass (Deimos) $\ll$ mass (Mars).

(iv) A digression on 'testing' and 'accepting' scientific theories
The formulation of the principle of empiricism raises some
further problems: namely, what do we mean here by 'testing'
and 'accepting'? Lakatos in his 1968 paper, 'Changes in the
Problem of Inductive Logic' (Lakatos, 1968), has subjected the
notion of 'acceptance of a scientific theory' to a very careful
analysis (p. 375 ff.) in which he distinguishes three senses of

acceptance (accept 1, 2 and 3). I will give here a slightly simplified account based on Lakatos' work.

First, all theories are accepted for certain practical purposes; particularly for the technological purpose of designing new machines. This notion of acceptance (called by Lakatos acceptance$_3$) is of course very important in general, but not so important for us as we are trying to give an analysis of theoretical science. Now in what sense, if any, might one accept a scientific theory for purely theoretical purposes? However often a theory has been tested, it should still be regarded with a certain measure of scepticism and subjected to further tests where possible. Despite this scepticism, there is, I think, a genuine sense in which scientific theories are accepted. They can be accepted (1) as worthy objects of further scientific study, and (2) as a basis for the design of instruments. I will deal with these points in turn.

Any scientific theory should, as I have said, be subjected to further tests. We should, moreover, attempt to explain it by deducing its approximate truth from some deeper theory. On the other hand it is not worth trying to explain or test any crankish idea. For example, it is not worth devising a theoretical explanation of the mechanism of telepathy because the existence of telepathy has not been sufficiently well corroborated to be an acceptable fact. In this sense we accept theories, which have been tested and corroborated, as worthy of further testing and of attempts at deeper explanation.

There is another important theoretical purpose for which theories can be accepted. We have seen that instruments are always designed in accordance with certain theories. Now which theories are we to accept for the purpose of designing instruments, or, more generally, for the purpose of determining the values of certain physical quantities? Here a genuine decision has to be taken. If the principle of empiricism formulated earlier is correct, we should only accept theories which have been checked against experience. Here, however, there is an apparent disagreement with Lakatos. Lakatos says (1968, p. 376): 'We may call acceptance$_1$ *"prior acceptance"* since it is prior to testing'; but for us no acceptance can be prior to testing.

This disagreement is more apparent than real for we are here using 'testing' in a different and wider sense than that employed by Lakatos. We have indeed employed a variety of expressions such as 'testing a scientific theory', 'checking it experimentally', 'matching it against experience', etc. All these are used in the following sense. We deduce from the theory (plus, of course, certain initial conditions and perhaps other assumptions) that a certain proposition should hold exactly or at least approximately. The proposition must be such that its truth or falsity can be or *has been* established experimentally. The theory passes the test if it gives the proposition the same truth-value as that assigned to it by experiment, and fails the test otherwise. Now here lies the difference with Lakatos. For Lakatos a test only occurs if some result is predicted which can be, but *has not yet been*, established experimentally. For a test we must derive some new result. Thus on Lakatos' position there can be experimental evidence for a theory which has never been tested. He goes on to say (1968, p. 376): 'We may call acceptance₁ *"prior acceptance"* since it is *prior* to testing. But it is *usually not prior to evidence*: most scientific theories are designed, at least partly, to solve an explanatory problem.'

We see then that the disagreement is indeed a verbal one. On the other hand this choice of a different terminology may involve a somewhat different attitude to certain questions. I would hold, for example, that a theory can be corroborated just as well by experimental results established before its invention as by new results established later and suggested by the theory. But could a new theory ever be accepted simply because it explained results already established experimentally and without deriving new results which were then shown to be true – without, in effect, being tested in Lakatos' sense? Newton's theory, as I have up to now expounded it, seems to be an example of this. The tests so far described all involve deriving from the theory results which had already been established. However, I have omitted several features of Newton's theory which should be examined.

Besides explaining Galileo's and Kepler's laws, Newton's theory also explained the laws of impact, the tides, the inequalities of the Moon's motions, and it enabled a theory of the figure of the Earth and of comets to be derived. Let us deal with these in turn. In the case of the laws of impact and the tides, Newton was explaining results which had already been established experimentally. Many of the inequalities of the Moon's motion had already been observed by astronomers, but Newton claims (1728, p. 578) that 'several irregularities of motion are deduced, not hitherto observed'. As regards the figure of the Earth, Newton deduced a flattening of the Earth at the poles. This was certainly a new result, but it was not checked experimentally till the expedition to Lapland of Maupertius and Clairaut. Moreover this expedition took place after Newton's theory had been generally accepted by the scientific world. The theory of comets constitutes the best example of genuinely new results which were not only derived from Newton's theory but also checked experimentally prior to the general acceptance of the theory. Newton's theory enabled the orbit of a comet to be calculated from a few astronomical readings. The predicted orbit of the comet could then be checked against its actual orbit as revealed by further astronomical readings. As Newton himself says (1728, p. 620):

> But the Problem may be resolved by determining, first, the hourly motion of a comet, at a given time, from three or more observations, and then deriving the orbit from this motion. And thus the determination of the orbit, depending on one observation, and its hourly motion at the time of this observation, will either confirm or disprove itself; *for the conclusion that is drawn from the motion only of an hour or two and a false hypothesis, will never agree with the motions of the comets from beginning to end.* [My italics]

Thus although the principal reasons for accepting Newton's theory at first were that it explained and unified a large number of results already established experimentally, it did predict some new results (certain inequalities of the Moon, and the paths of certain comets) which were shown experimentally to hold. Nonetheless I believe that Newton's theory would (or at least should) have been acceptable even if it had not been developed sufficiently to deal with the Moon's inequalities and comets. Although Newton's theory is not itself an example of this, it does show that a new theory might be acceptable even if the only evidence for it consisted of results which had been established experimentally before the introduction of the theory.

It might be objected that this account omits the important Duhem point that Newtonian theory contradicts the laws of Galileo and Kepler. Newton's theory predicts that the planets will move in ellipses with perturbations. The existence of the perturbations is a new result which can be checked empirically. Now it is true that Newton himself mentions one such perturbation. He says (1687, Book III, Prop. 13, Th. 13, p. 421): 'It is true, that the action of Jupiter upon Saturn is not to be neglected.... And hence arises a perturbation of the orbit of Saturn in every conjunction of this planet with Jupiter, so sensible, that astronomers are puzzled with it.' However, Newton here only explains a perturbation *qualitatively*, and moreover the perturbation had already been observed. The problem of calculating the perturbations is immensely complicated. Newton's theory would have had to have been accepted as 'worthy of such a complicated test' before the test could have been applied. Further, to predict the perturbations it is necessary to obtain values for the masses of the planets and the measurement of these values requires a use of Newtonian theory. I conclude that Newton's theory would have had to show itself to be highly acceptable before it could be used to deal, in detail, with the planetary perturbations.

(v) The problem of depth

This really concludes my account of conceptual innovation, but, before applying it to probability theory, I will mention three more general points in the philosophy of science which will prove important in the later discussion. The first of these is contained in Popper's paper 'The Aim of Science' (Popper, 1957b). Popper begins this paper by expounding his 'deductive' model of explanation. According to this account we explain an empirical law by deducing it from certain more general laws of nature plus, of course, initial conditions. If the explanation is a creative or original one, the empirical laws which are being explained will be familiar whereas the more general laws used to explain them will be newly invented. Thus (Popper, 1957b, p. 24) 'scientific explanation, whenever it is a discovery, will be *the explanation of the known by the unknown*'. These new unknown laws should be attempts to describe a deeper reality (1957b, p. 28):

...although I do not think that we can ever describe by our universal laws, an *ultimate* essence of the world, I do not doubt

that we may seek to probe deeper and deeper into the world
or, as we might say, into properties of the world that are more
and more essential.
   Every time we proceed to explain some conjectural law or
theory by a new conjectural theory of a higher degree of
universality, we are discovering more about the world, trying
to penetrate deeper into its secrets.

What does Popper mean by the concept of 'depth' used here? He
does not attempt to give a full account of the notion – of the
sense in which one theory can give a *deeper* description of reality
than another. He does however give a sufficient condition for
greater depth. This is illustrated by the historical example of
Newtonian mechanics.
   Popper begins his account by making the familiar Duhem
point that Newton's theory strictly speaking contradicts
Galileo's and Kepler's laws. Following Duhem, he argues that
this point goes against the inductive theory of the growth of
scientific knowledge. However his new idea is to apply this
point to the problem of depth. A higher-level theory is *deeper*
than the theories it explains if it *'corrects them while explaining
them'* (1957b, p. 10). In particular Newton's theory, because it
corrects Kepler's laws, is *deeper* than them. Of course it should
be emphasized that Popper's intention here is only to give *one*
sufficient condition for depth, and he himself says that there
may be others. One point should perhaps be made here. If we
accept the deductive model of explanation Newton's theory
does not strictly speaking explain Kepler's laws. From Newton's
theory plus initial conditions we do not deduce e.g. Kepler's
first law K that planets move in ellipses but only the modified
law K' that planets move approximately in ellipses or that
planets move in ellipses with small perturbations. We will retain
Popper's phrase that a deeper theory 'corrects a previous theory
while explaining it', but this must be interpreted to mean
'explains a modified version T' of the original theory T'. In this
way we can retain the strict deductive model of explanation.
   I will now give a second sufficient condition for depth, which
is a slight modification of Popper's but which will prove more
useful in the case of probability theory. Let us suppose that
before Newton we had not Kepler's law but only Schnorkelheim's

law S which stated that planets move round the Sun in closed curves which are vaguely though not exactly circular. Now Newton's theory is invented and from it we infer as usual S': planets move round the Sun in ellipses which are disturbed by small perturbations. Now S' does not contradict S in the way that K' contradicts K. In fact S' entails S. Thus S' does not correct S, but it does *render S more precise*. This 'rendering of a law more precise' also seems to me to be a sufficient condition for greater depth. Therefore I would alter Popper's slogan to: 'a new theory has greater depth than an older one if it *corrects* or *renders more precise* the older theory while explaining it'. As before, this slogan is to be interpreted in such a way that the deductive model of explanation can be retained.

(vi) The generalized principle of correspondence and related questions

I will next discuss two more problems raised by Popper. As is well known, he holds that we should value those scientific theories which have great generality and precision. He also believes that scientific theories are in no sense 'induced' from observations but are freely invented and then tested by experiment. Nonetheless he points out (1934, p. 276): 'one may discern something like a general direction in the evolution of physics – a direction from theories of a lower level of universality to theories of a higher level'. This phenomenon Popper calls 'quasi-inductive evolution'. It immediately creates a problem for his philosophy which he puts like this (1934, p. 277): 'The question may be raised: "Why not invent theories of the highest level of universality straight away? Is it not perhaps because there is after all an inductive element contained in it".' Strictly speaking this formulation is not quite right, since there is no 'highest level of universality', and thus we cannot reach it straight away.[1] However it does seem to be true that each new scientific theory goes only a little way beyond its predecessors, and we can ask whether there is some explanation for this in the logic of the situation.

A second closely related problem concerns the existence in scientific progress of what could be called a 'generalized principle

---

[1] I owe this point to T. Hook.

6

of correspondence'. Popper again values theories which are bold and which revolutionize our world picture as much as possible. On the other hand, it is nearly always the case that new scientific theories contain their predecessors as approximate cases. Popper is well aware of this and indeed says (1957b, p. 11): 'The demand that a new theory should contain the old one approximately, for appropriate values of the parameters of the new theory, may be called (following Bohr) the *"principle of correspondence".*' How is this conservative element to be explained in a philosophy which demands (for science though not for society): 'revolution in permanence'?

I have two reasons which explain the need for a generalized principle of correspondence. To begin with, the first test of a new theory will often be obtained by deriving the approximate truth of old results. This was the case with Newton's theory, and Newton's example has often been followed since. Suppose, however, that instead of using old results we derive some new ones and check these experimentally. Could we not employ this method on a totally new and revolutionary theory with no 'principle of correspondence'? I think not, because it must be remembered that the instruments we use to test the new theory will have been designed in accordance with the old. If the new theory totally contradicts the old, we will be deprived of instruments with which to test it.

A very nice example of this is provided by the famous test of Einstein's theory which was performed by making observations on the apparent displacement of stars at the time of a solar eclipse. Now Einstein's new cosmological theory involved the assumption that space is non-Euclidean. This was certainly a revolutionary step – one of the most revolutionary in the history of thought. On the other hand this new claim involved, as usual, a principle of correspondence: it was a consequence of the theory that Euclidean geometry held very well to a first approximation for terrestrial distances. I now claim that this correspondence was necessary because *the telescopes used to test the theory were designed in accordance with Euclidean geometry.* If the theory had completely contradicted Euclidean geometry, even for terrestrial distances, we would have been deprived of the use of telescopes in testing it. Of course certain tests would have been possible even without telescopes. One consequence of general relativity is that stones fall to the ground, and this is easily checked. On the other hand we must, I think, assume some kind of commensurability between the generality and precision of a theory and the severity and exactness of the tests to which it is subjected. But for exact tests to be possible we must retain a good deal of our old theoretical ideas.

Of course boldness, generality and precision are all excellent qualities in a new theory. There is, no doubt, a psychological limit on these imposed by the limited originality of human inventors. But, if the above argument is correct, there is also a limit imposed by the logic of the scientific situation. This solution is in fact very much the same as that offered by Popper himself, who says (1934, p. 277):

> Again and again suggestions are put forward – conjectures, of theories – of all possible levels of universality. Those theories which are on too high a level of universality, as it were (that is, too far removed from the level reached by the testable science of the day) give rise, perhaps, to a 'metaphysical system'....
>
> To obtain a picture or model of this quasi-inductive evolution of science, the various ideas and hypotheses might be visualized as particles suspended in a fluid. Testable science is the precipitation of these particles at the bottom of the vessel: they settle down in layers (of universality). The thickness of the deposit grows with the number of these layers, every new layer corresponding to a theory more universal than those beneath it. As the result of this process ideas previously floating in higher metaphysical regions may sometimes be reached by the growth of science, and thus make contact with it, and settle.

I have tried here to give a more detailed account of some of the features which cause a theory to be near the bottom of the vessel and so ready to deposit as science.

(vii) How the ideas of Part I will be applied to the theory of probability

Let us now finally comment on how the ideas of Part I are going to be applied to the theory of probability. In mechanics a set of empirical laws, notably Galileo's laws and Kepler's laws, were explained by a new deeper theory (Newton's theory) which involved the new and undefined concepts of 'force' and 'mass'. In probability theory, too, we have empirical laws: the Law of Stability of Statistical Frequencies and the Law of Excluded Gambling Systems. Our aim should be to exhibit probability theory as a theory which explains these laws in terms of the new concepts of 'probability' and 'independence'. Further, probability theory should not only explain these laws, but also correct them or render them more precise – thus proving itself to be a deeper theory. The new concept 'mass' did not acquire empirical significance through an operational definition but through the assumption that one mass (the mass of the Sun) was negligible in comparison with another (the mass of the Earth). If the analogy is going to hold here too, probability will not acquire empirical significance by means of a definition in terms of relative frequency (as von Mises claimed), but *through the decision to neglect one probability in comparison with another.*

In Part II a development of probability theory along the lines just indicated will be attempted.

PART TWO

The Axiomatic Superstructure

CHAPTER FOUR

Probability and Frequency

The aim of this part is to set up the axioms of probability in such a way that from these axioms (and perhaps certain other principles) we can deduce the two empirical laws of probability theory: the law of stability of statistical frequencies, and the law of excluded gambling systems. It is obvious that this deduction cannot be carried out from the ordinary Kolmogorov axioms *alone*, for these axioms present probability theory as a formal mathematical system (the study of probability spaces $(\Omega, \mathfrak{F}, p)$) and do not show how this theory is to be tied in with experience. Normally the question of how the abstract calculus is connected with experimental results is treated in some supplementary informal remarks; but I will try to deal with it in a more formal and explicit manner. This task is closely connected with certain more traditional problems. I will start by discussing these problems, and this will lead to my main theme. The advantage of such a procedure is that it enables me to relate my approach to previous work.

(i) Three views on the relations between probability and frequency

In his 1933 monograph on the 'Foundations of the Theory of Probability' Kolmogorov devotes only one short, but nonetheless very interesting section (§ 2), to the problem of how the axioms are related to experience. He says (1933, p. 3) 'There is assumed a complex of conditions, $\mathfrak{S}$, which allows of any number of repetitions.' The set Ω (which is the first member of the probability space $(\Omega, \mathfrak{F}, p)$) is then the set of all possible outcomes of $\mathfrak{S}$ 'which we regard *a priori* as possible'. More precisely the points of Ω are outcomes of $\mathfrak{S}$ as completely specified as we require. Such completely specified outcomes Kolmogorov calls 'elementary events'. Sets of elementary events, i.e. subsets of Ω, give possible outcomes of $\mathfrak{S}$ in a more generalized sense. Thus if $\mathfrak{S}$

specifies the repeatable conditions of throwing a certain die, then a completely specified outcome would be '1' or '2' or...or '6'. These are the 'elementary events'. A generalized outcome might be 'the result was odd'. This event is the subset consisting of elementary events '1', '3', '5'. Of course what we regard as a 'complete specification of the outcome' can vary with the circumstances. If we were not interested in the exact number obtained, but only in whether it was divisible by two, we could take 'even' and 'odd' as our elementary events.

Consider now an event A ($\subseteq \Omega$). Suppose $\mathfrak{S}$ is repeated n times and A occurs $m(A)$ times. Everyone would agree that there is a close relation between $p(A)$ and $m(A)/n$. But what exactly is this relation, and how is it established? This is the classic problem of probability and frequency. To indicate the variety of possible answers to this question I shall now list three viewpoints all adopted at one time by eminent probabilists, viz. von Mises, Doob and Kolmogorov.

Von Mises' view is so familiar that it is hardly worth stating again. It is that $p(A)$ and $m(A)/n$ are connected by the equation

$$p(A) = \lim_{n\to\infty} m(A)/n$$

and that this equation constitutes the definition of $p(A)$. The only point worth adding here concerns the question of how von Mises altered his view in the light of the Kolmogorov axioms. In fact as we shall see he adopted successively two different positions. First he claimed that Kolmogorov's axioms were *consistent* with his frequency theory. Then, more radically, he argued that Kolmogorov's axioms could not be reconciled with the frequency view, and so should be abandoned. The first of these approaches was adopted in his symposium with Doob on the foundations of probability in 1941. Here he says that his limiting frequency definition of probability is a necessary supplement to the Kolmogorov axioms, needed in order to establish the connection with experience. In his own words (1941, p. 347): 'All axioms of Kolmogoroff can be accepted within the framework of our theory as a part of it but in no way as a substitute for our definition.'

Doob, on the other hand, claims that the connection between frequency and probability is established by certain theorems of the probability calculus (1941, p. 208):

The justification of the above correspondence between events and Ω-sets is that certain mathematical theorems can be proved, filling out a picture on the mathematical side which seems to be an approximation to reality, or rather an abstraction of reality, close enough to the real picture to be helpful in prescribing rules of statistical procedure.

In particular he singles out the strong law of large numbers which he calls theorem A. Actually he considers the following particular case. Suppose we have a sequence of independent trials whose results are expressed by whole numbers between 1 and N inclusive. Suppose result j has probability p_j and that in the first n trials it occurs n_j times, then

$$\text{prob}\left(\lim_{n\to\infty} n_j/n = p_j\right) = 1.$$

Doob says (1941, p. 208), 'Theorem A corresponds to certain observed facts relating to the clustering of "success ratios" giving rise to empirical numbers $\bar{p}_j$'; and again (1941, p. 209), 'In all cases, such mathematical theorems as Theorem A...give the basis for applying the formal apparatus in practice.' Thus we do not have to define probability as limiting frequency. The equality of probability and limiting frequency is 'derived' (in some sense) from the strong law of large numbers.

Kolmogorov differs from both von Mises and Doob in that he considers not an infinite sequence of repetitions but merely a 'large number'. He says (1933, p. 4):

Under certain conditions, which we shall not discuss here, we may assume that to an event A which may or may not occur under conditions $\mathfrak{S}$, is assigned a real number $p(A)$, which has the following characteristics:

(a) One can be practically certain that if the complex of conditions $\mathfrak{S}$ is repeated a large number of times n, then if m be the number of occurrences of event A, the ratio m/n will differ very slightly from $p(A)$.

(b) If $p(A)$ is very small, one can be practically certain that when conditions $\mathfrak{S}$ are realized only once the event A would not occur at all.

A number of obvious objections can be put forward to this view. How large in (a) is a 'large number' to be? How small in (b) is 'very small'? Also, if one translates the phrase 'one can be practically certain' into 'there is a high probability that' a circularity appears to manifest itself. On the other hand, as against Doob and von Mises Kolmogorov is here closer to experience in that he considers only a finite number of repetitions; and certainly one can never have an infinite number of repetitions in practice. It could be held (though I do not agree with this view) that the vagueness of his formulations is unavoidable because the relation to experience is essentially an imprecise matter.

One result is worth mentioning. It is a consequence of Kolmogorov's position that if we repeat the conditions $\mathfrak{S}$ a large number n of times, then, with the usual notation, the equation

$$m(A)/n \doteq p(A) \qquad (1)$$

holds, or at least holds with high probability. I think everyone would agree with the truth of (1). Indeed a probability theory in which we did not have (1) would be extraordinary and most unacceptable. This may seem a trivial result but I shall use it later to argue for one of my proposals.

I do not want to claim that these three accounts exhaust the possible ways in which probability can be related to frequency. Indeed the theory which I will develop later, though it embodies elements from each of the three approaches, is slightly different from all of them. These different views suggest, however, a way of classifying philosophical theories of probability. I have already divided such theories into 'scientific' and 'logical'. I now propose subdividing scientific theories according to the account they give of the relation between probability and frequency. More precisely let us say that a scientific theory of probability is a *frequency* theory if it identifies probabilities with relative frequencies or limiting (relative) frequencies *either* in the mathematical formalism *or* as an additional informal postulate designed to tie the axioms in with experience. According to this account Kolmogorov's theory of probability is a frequency theory. This is in agreement with what Cramér (who by and large follows Kolmogorov on this point) says (1945, p. 148):

Whenever we say that the probability of an event E w.r.t. and experiment $\mathfrak{E}$ is equal to p, the concrete meaning of this assertion will thus simply be the following: In a long series of repetitions of $\mathfrak{E}$, it is practically certain that the frequency of

E will be approximately equal to p. – This statement will be referred to as the frequency interpretation of the probability p. [In the original this whole passage is in italics]

On the other hand Doob's position quoted above is a scientific but non-frequency theory of probability, for the relation between probability and frequency is established by means of a theorem of the calculus (the law of large numbers). Our own view is also a scientific but non-frequency one.

This classification does, however, have one curious feature. Popper's propensity theory (to be discussed in Chapter 7) turns out to be a frequency theory (in our sense), although Popper claims to have abandoned the frequency theory in adopting the propensity view (but see footnote on p. 148). Such differences in terminology are really inevitable. Like all classifications, the one given above depends very much on the standpoint of its author. From different theoretical positions different features will seem of greater or of less importance, and classification will accordingly differ. One has really to decide which is the best theory before one can decide which is the best classification.

Our problems are concerned with the relations between *probability spaces* $(\Omega, \mathfrak{F}, p)$ and the repeatable conditions $\mathfrak{S}$ on which these are defined. They cannot therefore be discussed within the standard formalism of probability theory which only treats of probability spaces and leaves the conditions $\mathfrak{S}$ to be introduced in certain informal explanations appended to the account. In order to get over this difficulty I propose to introduce the notion of a *probability system* which will be defined as an ordered quadruple $(\mathfrak{S}, \Omega, \mathfrak{F}, p)$ where $(\Omega, \mathfrak{F}, p)$ is a probability space and $\mathfrak{S}$ is a set of conditions s.t. Ω is the set of *a priori* possible outcomes of $\mathfrak{S}$. In terms of this notion it will be possible to state our problems and proposed solutions with some degree of precision. The Kolmogorov axioms in effect state that p is a non-negative completely additive set function defined on Ω s.t. $p(\Omega) = 1$. In addition to these I will propose another axiom (to be called the axiom of independent repetitions) for probability systems. As a preliminary to this, I will next compare in detail von Mises' mathematical development of probability theory with Kolmogorov's. This comparison will help to explain the motivation behind the new axiom.

(ii) Randomness in von Mises' theory

It will be remembered (pp. 3–7) that the basic notion of von Mises' system is that of the *collective*. In fact it was pointed out that there is an ambiguity in his use of the term. It sometimes denotes the long but necessarily finite sequences of events which actually occur, e.g. long sequences of tosses of a coin, of men who have insured themselves with a certain company, of molecules of a gas. At other times it refers to the infinite sequences of attributes which are introduced as mathematical abstractions of the finite sequences. It was decided that we would only use 'collective' to refer to the infinite sequences of the mathematical theory, and would call the actually occurring finite sequences 'empirical collectives'. Empirical collectives satisfy the familiar two laws and corresponding to these we introduce two axioms for (mathematical) collectives. The first of these (the axiom of convergence) presents no problems. However, there are difficulties with the formulation of the second (the axiom of randomness). This question will now be discussed.

The empirical law states that it is impossible to improve one's chances by using a gambling system. Now in a collective $\mathfrak{C}$, the probability of an attribute $A(p(A)) = \lim_{n \to \infty} m(A)/n$. So the axiom of randomness becomes (roughly) that in any subsequence obtained from the original collective by means of a *place selection*, $m(A)/n$ must continue to converge to the original value. The trouble is that this axiom appears to be contradictory; or more strictly to render the class of collectives empty except in the trivial case when the probability of each attribute is either 0 or 1. For suppose that attribute A has a probability $p(A)$ greater than zero and less than one. By the first condition A must appear an infinite number of times. Thus we can choose a subsequence consisting just of attributes A. For this subsequence we have $\lim_{n \to \infty} m(A)/n = 1 \neq p(A)$, and so the axiom of randomness is contradicted. The way out of this difficulty is to restrict the class γ of place selections or gambling systems in such a way that this unpleasant consequence is avoided. The problem is how to do this.

Von Mises himself suggested that we make the following stipulation (1928, p. 25): 'the question whether or not a certain member of the original sequence belongs to the selected partial sequence (is)...settled *independently of the result* of the corresponding observation, i.e. before anything is known about this result'. However, this is not very satisfactory because it gives an epistemological definition of an allowable selection whereas we really need a mathematical one. This desire for a mathematical definition doesn't arise merely from a mania for precision. Only if we have a mathematical definition of the class of allowable place selection or gambling systems will we be able to prove the existence of collectives. Such an existence proof is highly desirable to defend the notion of collective against charges of inconsistency, for, in addition to the inconsistency proof just given, another putative such proof was formulated.

The proponents of this objection, Cantelli (1935, §§ 7, 10, 12) and T. C. Fry (1928, pp. 88–91), claim that there is a contradiction between the limiting frequency definition of probability and the binomial formula, i.e. the formula which states that the probability of getting the event A (probability p) r times out of n independent trials is $^nC_r p^r (1-p)^{n-r}$. However, the binomial formula is of course derivable in von Mises' theory. So if this objection is correct his theory must be inconsistent. The argument is this. Suppose we have a repeatable experiment E (say the tossing of a coin), and as usual the probability of outcome A is p. Obtain a sequence S by repeating E indefinitely. According to von Mises $p =$ the limiting frequency of A in S. So, given $\epsilon > 0$ there must be an N such that the difference between p and the relative frequency of A is less than ϵ for all $n > N$. But let us now consider any finite segment of the sequence immediately following the first N elements (say the elements $N + 1, \ldots, N + m$). According to the binomial formula there is a finite probability of getting A at each of these elements, viz. p^m. If we get a run of such successes long enough, the relative frequency of A will diverge from p by more than ϵ for any sufficiently small ϵ. There is thus for any N a finite probability of such a divergence contrary to the requirements of the limit definition.

To resolve this difficulty we need only consider the meaning of the assertion that there is a probability p^m of getting A at the $N + 1, \ldots, N + m$ places of S. According to von Mises the meaning is this. If we produce an infinite number of sequences $S', S'', \ldots, S^{(n)}, \ldots$ in the same way as S, the limiting frequency of those which have A at the $N + 1$th $\ldots N + m$th places will be p^m. This is not at all incompatible with the sequence S quite definitely *not* having A at all these places. We would only get a contradiction if we postulated not only that the relative frequency of A converged to p for each $S^{(i)}$ but also that the convergence was uniform over the $S^{(i)}$.

This I think satisfactorily solves the difficulty, but even if this particular solution is not correct, an existence proof of the kind specified would vindicate the frequency theory. Suppose it were shown that there are sequences which both satisfy the limiting postulate and for which the binomial formula is derivable. The alleged contradiction between the two requirements is thereby shown to be only apparent, and an advocate of a theory of von Mises' type need not worry about it.

These considerations enable us to formulate the problem of randomness as follows: we have to specify a set of allowable selections satisfying three conditions. First, we must make the set of selections large enough to enable the probability calculus to be developed. For example, we must be able to obtain the multiplication law using our definition. Secondly, the set must be small enough to prove that collectives relative to the set do in fact exist. Lastly, we do not want this set of selections to be entirely arbitrary, but rather to correspond as closely as possible to von Mises' intuitive idea of excluding all gambling systems.

In the period 1919–40 this problem of randomness attracted a great deal of attention. A complete list of those who contributed to it would have to include the names of Church, Copeland, Dörge, Feller, Kamke, von Mises, Popper, Reichenbach, Tornier, Waismann and Wald. I will not attempt a complete history but rather concentrate on the work of Wald and Church whose combined efforts constitute, in my opinion, a complete and satisfactory solution to the original problem.

In this treatment I will also make a couple of further simplifications. Let us suppose our putative collective $\mathfrak{C}$ has attribute space M. Then Wald considers the question of defining not only the set γ of allowable gambling systems but also the subsets $\mathfrak{M}$ of M for which we introduce probabilities. If M is an infinite set, we cannot in general take $\mathfrak{M}$ to be the set of all subsets of M and thus we have to decide how $\mathfrak{M}$ has to be chosen. As we are more interested in the question of gambling systems we will omit this consideration. If M is finite, then $\mathfrak{M}$ can be taken to be the set of all subsets of M, and the problem does not arise in this case. A second question concerns the existence proof for collectives. For many definitions of the set γ of allowable gambling systems an existence proof can be given, but it is non-constructive. It would not therefore be acceptable to intuitionists who regard only constructive existence proofs as valid. Many of the authors mentioned (including Wald and Church) have taken this objection seriously and have considered modifications of their definition of γ which would render a constructive existence proof possible. I myself feel that these intuitionistic scruples are not really justified. After all, elsewhere in applied mathematics the whole body of classical mathematics is assumed without question, and I cannot see any need for introducing higher standards of rigour in the case of probability theory.

Wald's results are contained in his 1937 paper *Die Wider-spruchsfreiheit des Kollektivbegriffes*. The same results (but without proofs) were again given in German in 1938 in a shorter paper of the same title now reprinted in his selected papers. There is also a summary in English of his work on this problem in von Mises (1963, pp. 39–43). Bearing in mind the qualifications of the previous paragraph, we can state Wald's main theorem very simply. If we confine ourselves to a denumerable class γ of place selections or gambling systems, then there exists a continuum infinite set of collectives or random sequences having any assigned probability distribution. So, far from random sequences being rare or even non-existent, they are much more numerous than sequences which exhibit regularity. There is still a certain arbitrariness left in the choice of γ. Wald tries to mitigate this by two considerations. First of all, in any particular problem we certainly won't want to consider more than a denumerable set of gambling systems. Secondly, let us suppose we are formulating our theory within some logical system, e.g. (to quote his example) *Principia Mathematica*. Within such a system we only have a denumerable set of formulas and so can only define a denumerable set of mathematical rules.

This last remark of Wald's may have suggested to Church a way in which the set γ could be specified more precisely. He gives his idea in the short paper 'On the Concept of a Random Sequence' (1940). After stating the familiar objection to the existence of collectives he goes on to say: 'A *Spielsystem*[1] should be represented mathematically, not as a function, or even as a definition of a function, but as an effective algorithm for the calculation of the values of a function.' This point, once made, must surely be recognized as correct. A gambling system after all is nothing more than a rule telling us at each throw whether to bet or not. Such a rule must deliver its instructions in a finite time, in other words it must be an effective procedure for determining whether we are to bet or not. If we accept Church's thesis that recursiveness is an adequate mathematical replacement for the notion of effectiveness, we arrive at the following definition of a gambling system.

Let $a_1, a_2, \ldots, a_n, \ldots$ be our original sequence. We will suppose for simplicity that it is composed of 0s and 1s. Then the sequence $c_1, c_2, \ldots, c_n, \ldots$ of 0s and 1s is a gambling system if it is defined in the following manner:

$$c_n = \phi(b_n)$$

where (i) $b_1 = 1$, $b_{n+1} = \psi(a_n, b_n)$, (ii) ϕ, ψ are recursive functions of positive integers and pairs of positive integers respectively; and if the integers n such that $c_n = 1$ are infinite in number. The introduction of the b_i in (i) is merely a device to ensure that our decision whether or not to choose a (say) can depend on the preceding members of the a-sequence as well as on n. We use this gambling system by selecting only those members a for which $c_i = 1$. Since there are only a denumerable number of recursive functions, Wald's theorem shows that there is a continuum infinity of such 'recursively random' sequences with any assigned distribution of probabilities.

Church's notion of a recursively random sequence and Wald's existence proof put von Mises' theory on a thoroughly satisfactory mathematical basis and invalidate any claim that the theory is inconsistent. However, it will I think be interesting to mention some further work on this subject which has appeared since 1940. In fact at least two new ideas have been suggested.

The first of these was proposed by Loveland (see his paper, 1966). He extends Church's definition by allowing place selections for which the decision to choose or reject a_n depends

---

[1] i.e. gambling system.

effectively not only on n and a_i for $i < n$, but also on certain
values a_i with $i > n$. He shows that there are random sequences
in Church's sense which are not random in his stronger sense.
A second development is a definition, due to Kolmogorov, of
randomness for finite sequences. We begin by introducing a
certain class of 'universal algorithms', and define the complexity
of a finite sequence as the length of the shortest universal
algorithm which computes it. We define a random (finite)
sequence to be one of maximal complexity. This definition can
be extended to the case of infinite sequences where it largely
agrees with previous definitions of randomness. This extension
is carried out by Martin-Löf in his paper (1966), which gives a
good account of this particular line of thought.

Despite the interest of these developments, they will not be
discussed further. My aim rather is to raise the following question.
Do these 'frequency theory' investigations have any relevance
within the modern measure theoretic approach to probability?
After all, it is possible to develop the theory of probability and
even of statistics without mentioning anything like the Church
definition of randomness. This is done, for example, by Cramér
(1945). Is the Church definition therefore redundant or can we
still accord it some importance? I shall discuss this problem in
Chapter 6. For the moment the mention of measure-theoretic
approaches is a convenient reminder that we must return to
our comparison between von Mises and Kolmogorov.

(iii) Comparison between von Mises and Kolmogorov

So far we can observe the following differences between these
two authors. Von Mises is prepared to apply the probability
calculus to empirical collectives; Kolmogorov to certain repeatable conditions. In fact the difference here is not as great as it
might seem at first sight. If we repeat a set of conditions a large
number of times, we might well get an empirical collective (in
the sense of von Mises). Such empirical collectives could be
looked upon as the repeatable conditions viewed in an exten-
sional manner. However, even with this method of harmonizing
there still remain important differences. First of all, not every
empirical collective (i.e. not every long sequence of events
satisfying the two empirical laws stated above) is produced by
the continual repetition of a certain uniform group of conditions.

7

(I will give a counter-example proposed by Popper later.) In this respect von Mises' notion is wider. Again, von Mises will only allow the introduction of probabilities in physical situations where there is an empirical collective – where, therefore, we have a long sequence of observations. But if we accept Kolmogorov's account, there is nothing to prevent us introducing probabilities in a situation where we have repeatable conditions – even though these repeatable conditions may only have been repeated a few times, perhaps indeed only once. In this respect Kolmogorov's account is the wider and appears better suited to the introduction of probabilities for single events.[1] A second difference is the obvious one that von Mises defines probability in terms of other notions whereas Kolmogorov takes it as a primitive concept. It must be added by way of qualification that Kolmogorov has to introduce an informal explanation of how probability and frequency are to be identified – but still an important divergence remains. Lastly, and perhaps most surprising of all, we must note that there is nothing in Kolmogorov's system corresponding to von Mises' axiom of randomness. In the previous section I have discussed at great length the efforts that were devoted by von Mises and his followers to the problem of randomness. Yet no mention of the matter appears in Kolmogorov's account. In a sense the next chapter can be considered as an attempt to introduce randomness into the Kolmogorov axioms.

To sum up, then, we have noted three differences between von Mises and Kolmogorov:

(1) von Mises introduces probabilities for the attributes of empirical collectives; Kolmogorov for the outcomes of certain repeatable conditions.
(2) von Mises defines probability in terms of frequency; Kolmogorov takes it as a primitive notion.
(3) There is nothing in Kolmogorov's system corresponding to the axiom of randomness.

It should be stressed that these are not the only differences between the two authors but rather the differences which will concern us in what follows.

---

[1] This last point is amplified in Chapter 7.

(iv) Two criticisms of Kolmogorov by von Mises
Von Mises summed up his differences with Kolmogorov by
claiming that while Kolmogorov considered only the mathematical developments, he was interested in producing a founda-
tion for the theory which would exhibit its relation to experience.
He claimed, for instance (1928, p. 99): 'Our presentation of the
foundations of probability aims at clarifying that side of the
problem which is left aside in the formalist mathematical
conception.' I agree with von Mises that there remain problems
about the relation between theory and experience, and I am in
effect following him in trying to deal with these problems. I
disagree, however, with his attempted solution of these problems.
In particular, it seems to be possible to give an adequate
account of the matter without diverging very far from Kolmogorov's methods and concepts. It is not necessary, for example,
to define probability in terms of frequency or to use the special
concepts of randomness and place selection. My object is thus
to produce a supplement to Kolmogorov's axioms, a supplement
which will clarify the relations between theory and experience
but which will be quite consistent with Kolmogorov's system
and so need not affect mathematicians working at an abstract
level.

This programme should be contrasted with von Mises' final
position set out in *Mathematical Theory of Probability and
Statistics*, 1963. Here von Mises presented his frequency theory
not as a necessary addition to the Kolmogorov axioms (his 1941
view) but as a competing mathematical theory which diverged
in a number of respects from Kolmogorov's and was preferable
where it did diverge. Von Mises had in fact two main arguments
against Kolmogorov which we will now summarize and attempt
to refute.

His first objection concerns Kolmogorov's definition of
independence for two events, i.e. two sets A and B of $\mathfrak{F}$. These
are said to be independent if the multiplication law $p(A \cap B) =
p(A)p(B)$ is satisfied. Von Mises objects that this definition
leads to calling two events independent which are not independent in the intuitive sense of 'being separate' and 'not influencing
each other'. The multiplication law can be satisfied due to
'purely numerical accidents' (1963, p. 37), even though the
intuitive requirements are not met. Consider, for example,

$\Omega = (1, 2, 3, 4)$, $p(1) = p(2) = p(3) = p(4) = 1/4$, $A = (1, 2)$, $B = (2, 3)$. A and B are independent according to Kolmogorov's definition but not in the intuitive sense. Von Mises continues with the following piece of rhetoric (1963, p. 38):

> In a meaningful concept of independence *two* 'properties' are involved which may or may not influence each other. In contrast to that, a definition based on Eq. (38)[1] (to which we did our best to give a meaningful semblance) remains a watered-down generalization of a meaningful concept.

Von Mises' argument takes the following form. We start with a certain assumption (in this case Kolmogorov's definition of independence). From this a result is derived which is considered to be 'an intuitive contradiction'. It is then concluded that the original assumption must be abandoned. Now such arguments involving 'intuitive' contradictions are by no means always valid. Consider, for example, the Russell–Frege definition of an infinite cardinal: this leads to the result that there are the same number of even integers as integers, a result which goes against the intuitive requirement that the number of integers should be twice that of the even integers. (This 'contradiction' was in fact noticed by Leibniz.) However, we should now have no hesitation in adopting the Russell–Frege definition and abandoning the intuitive requirement.

On the other hand, I don't think that arguments using intuitive contradictions are always invalid. To see what weight such an argument might have let us consider what these 'intuitions' amount to. They are, of course, nothing but our earlier theories about the subject – perhaps unconsciously held and never explicitly formulated. Now we do not always want to retain our older theories, our 'intuitions', but there is still a certain weight which we must attach to them. This is nothing but the 'principle of correspondence' again, and I have argued earlier that there is a need for such a principle. What it amounts to is this. We cannot allow a new theory arbitrarily to contradict our ordinary presuppositions. If it does so, there must be good reasons given.

Well then, when is an argument involving an intuitive contradiction valid and when is it invalid? This is a delicate matter

---

[1] The multiplication law.

which must be judged in individual cases. No general rule can be given. Broadly though, if the new theory is very powerful and elegant whereas any theory embodying our old ideas would be clumsy and unmanageable then we will prefer the new theory. This covers the Frege–Russell case. If on the other hand the intuitive contradiction can be avoided in theory A, which is every bit as satisfactory on other points as theory B, then the contradiction gives us good reason for preferring A to B. This, I hope, will be the situation when we use arguments involving intuitive contradictions below (Chapters 5 and 7). The present case is, I believe, of the first kind. Kolmogorov's definition of independence is so simple and elegant that we should be prepared to tolerate the oddity which von Mises has brought to light.

Von Mises' second objection is along the same lines but is more interesting, both in itself and for the light which it throws on his philosophical presuppositions. According to him we should only introduce such probabilities as are empirically verifiable at least in principle. As he puts it (1963, p. 47): '*Our* aim is to build up a coherent theory of verifiable events.'

Now unfortunately no probability statement within von Mises' theory is strictly verifiable – for a strict verification would involve going through a complete infinite collective and calculating limits. However, by a 'verification' von Mises means such approximate verifications as can be obtained by considering very long sequences – empirical collectives in fact. Thus (1963, p. 71) he considers 'the probability that in the continued throwing of a die the "six" will never appear'. This probability is of course zero in the usual case – a fact which can be 'verified' as follows (1963, p. 72): 'If we repeatedly make $n = 1,000$ trials there will be at least one six in the overwhelming majority of these groups of 1,000 trials each, and if $n = 10,000$ this will hold in an even more striking way.' Even with this liberal notion of verification certain probabilities cannot be verified. An example given (1963, p. 82) is: 'the probability...that in the continued tossing of a die, 6 appears a finite number of times only'. Whether we take $n = 1,000$ or $n = 10,000$ or etc., we can never be even approximately sure if we have an instance of this event occurring or of its not occurring. Therefore the corresponding probability is not verifiable – even in principle. Suppose, however, that in Kolmogorov's system we take as basic set Ω the set of sequences formed from the numbers 0, 1, ..., 6. By regarding these sequences as expansions to the base 7 we can look on Ω as the unit interval $[0, 1]$, and take $\mathfrak{F}$ to be the Borel sets of $[0, 1]$. This would be a fairly standard way of tackling the problem. However, the set of sequences which involve 6 only a finite number of times is a Borel set, and so would be assigned a probability.

To escape this conclusion von Mises set himself the task of devising a field of subsets of $[0, 1]$ which would contain just those sets which corresponded to verifiable probabilities. His solution was to consider the field F_1 of sets which (1963, p. 74) 'have content in the sense of Peano and Jordan'.[1] These can be characterized as measurable sets whose boundary has measure zero. Intuitively they can be equated with verifiable events for the following reason. If the result lies on the boundary of the set we cannot be sure whether it is an instance of the event or not. If the boundary has measure zero, this uncertainty will not affect the probability, but if the boundary has non-zero measure it will. F_1 is a subset of the measurable sets and overlaps the Borel sets. It has a higher cardinality than the Borel sets but is a field rather than a σ-field. It should be noted that when we talk here of measure it is assumed that a probability measure has been introduced on the intervals of $[0, 1]$. This can be extended to give a Lebesgue–Stieltjes measure and we speak of sets being measurable w.r.t. this measure. Of course von Mises does not want correspondingly to extend the ascription of probabilities to all these sets but only to a subset of them, namely the set F_1.

One has to admit that von Mises' solution of his problem is most ingenious but unfortunately his whole task is misconceived. The verifiability criterion has been criticized so damagingly by Popper that no serious philosopher of science would adopt it today. If we require that every statement of a scientific theory should be 'verifiable in principle', then we exclude for example all universal laws. This is really sufficient in itself to dispose of the criterion. What we should require instead, following Popper, is that a scientific theory *as a whole* should be falsifiable or testable. Now admittedly there are theoretical difficulties about the falsifiability or testability of probability statements, but in

---

[1] The use of sets of this character in probability theory had already been suggested by Tornier and Wald, cf. their papers cited in the list of references.

practice statisticians certainly do devise and carry out tests. The mathematical theory of such tests can be developed in a framework in which probabilities are freely assigned to all Borel sets. (Such an exposition is made in Cramér's book, 1945.) Thus although this objection of von Mises would have had some force within a strictly positivistic philosophy, it has none within a present-day Popperian account of scientific method.

CHAPTER FIVE

Repeatability and Independence

(i) Analysis of repeatability

We have accepted Kolmogorov's suggestion that probabilities should be defined for the outcomes of sets of conditions which 'allow an indefinite number of repetitions' or which (more briefly) are 'repeatable'.

Granted this assumption our aim is to investigate rather more closely than is usually done the notion of 'repeatability' thereby introduced; and, in particular, to investigate its connection with the notion of 'independence'. One view of the matter is that repeatability entails independence. This is adopted by Popper who writes (1957a, referring to the equation '$p(a,b) = r$'):

> The objectivist view of b is that b states the repeatable conditions of a repeatable experiment. The results of previous experiments are no part of the information b; and if they were part of b, they could be omitted without loss; for they must be irrelevant, since repeatability *entails* independence from predecessors: since otherwise the new experiment would not repeat the conditions of the previous one.

I will argue, however, that repeatability does *not* entail independence, and will discuss some implications of this result.

We will begin with the obvious point that any two alleged repetitions will be found on closer inspection to differ in many respects. Consider for example two tosses of a coin which would ordinarily be regarded as repetitions. Closer inspection might reveal that in one case the head had been uppermost before the toss was made and in the other the tail. Moreover, even if every macroscopic property of the tossing procedure did appear to be the same in the two cases there would still be the difference that the two tosses occurred at different times. We must therefore regard two events as repetitions not if they are the same in every way (which is impossible) but if they are the same in a certain well specified set of ways. Two events are not in themselves repetitions. The question of whether they are such depends on how we are proposing to describe them. Any two events, however similar, will differ in certain respects and this could bar them from being repetitions. Conversely, any two events, however dissimilar, will agree in certain respects and this could lead to their being considered as repetitions. In fact a sequence is a sequence of repetitions only relative to some set of common properties or conditions. These considerations suggest the following definition. A sequence is a sequence of repetitions relative to a set $\mathfrak{S}$ of conditions or properties if each member of the sequence satisfies every condition of $\mathfrak{S}$, and *irrespective of how the members differ in other respects*. This definition is all right as far as it goes, but it will be convenient to modify it in the light of another aspect of the matter.

In any sequence of repetitions there is characteristically not only a set of constant features but also some variable feature. Usually this variable parameter is time, as in the case of a sequence of tosses of a coin, but this need not be so. Consider, for example, twenty students carrying out the 'same' experiment at the same time. Here the variable parameter is spatial position. Yet again the variable parameter may include both spatial and temporal components. This leads us to the following difficulty. Consider the case of the twenty students and suppose that they are performing an electrical experiment. We might not want to consider such experiments as repetitions even though some set of defining conditions were satisfied if, in addition, the pieces of apparatus were so close together that some kind of magnetic interference occurred. We would require in effect that the experiments should be sufficiently widely spaced in position. Similarly, in the temporal case we might want the various events to be sufficiently widely spaced in time. Of course we could regard this matter as being included in the relevant set of conditions $\mathfrak{S}$, but I think it will be better to treat it separately by introducing the concept of a *spacing condition*. We will henceforth specify that any sequence of repetitions must involve a spacing condition s stating that the elements of the sequence must be separated in such and such a way relative to some variable parameter (e.g. time or spatial position or a combination of both). This proves convenient because we sometimes want to consider the same basic set of conditions $\mathfrak{S}$, but different spacing parameters (s, u, v say). We may also mention that although the examples of spacing parameters given so far involve space and time, this need not be so. The parameter may have a more abstract nature. Further, it may specify certain initial conditions which the first of the sequence of repetitions must satisfy. Examples will be given later.

We can now reformulate our definition of 'sequence of repetitions' as follows. A sequence of events is a sequence of repetitions relative to a set of conditions $\mathfrak{S}_s$ which includes a spacing condition s, provided that all the conditions $\mathfrak{S}$ are satisfied by each event, and the events are spaced as required by s. A set $\mathfrak{S}_s$ of conditions is repeatable provided that an indefinitely long sequence of repetitions relative to $\mathfrak{S}_s$ is in principle (i.e. as far as the general laws of nature are concerned) possible.

We can now discuss the question of combining a number of different repeatable conditions. Suppose we have n different repeatable conditions $(\mathfrak{S}_1)_s \ldots (\mathfrak{S}_n)_s$ which all have the same spacing condition s. We can then consider the new condition which is satisfied by an ordered sequence $(e_1, \ldots, e_n)$ s.t. e_1 satisfied $\mathfrak{S}_1, \ldots e_n$ satisfies $\mathfrak{S}_n$ and $e_1 \ldots e_n$ are spaced in accordance with s. For this new condition we can specify a new spacing parameter s' which may or may not be the same as s. The new condition + spacing parameter we write $(\mathfrak{S}_{1s} \wedge \ldots \wedge \mathfrak{S}_{ns})_{s'}$. Note that $(\mathfrak{S}_{1s} \wedge \ldots \wedge \mathfrak{S}_{ns})_{s'}$ need not be a repeatable condition. The occurrence of $\mathfrak{S}_1$ may exclude the occurrence of $\mathfrak{S}_2$ at a subsequent event spaced according to s. In this case $(\mathfrak{S}_{1s} \wedge \ldots \mathfrak{S}_{ns})_{s'}$ could not in logic be instantiated at all.

Example: Let $\mathfrak{S}_1$ be '...is a Monday', $\mathfrak{S}_2$ be '...is a Tuesday' etc. Let s be 'two successive events have not more than six days in between them'. Let s' be 'two successive events are separated by at least a year'. Then a sequence of repetitions of $(\mathfrak{S}_2)_s$ is a sequence of successive Tuesdays; a sequence of repetitions of $(\mathfrak{S}_{1s} \wedge \ldots \mathfrak{S}_{7s})_{s'}$ is a sequence of weeks; successive members being more than a year apart.

One special case will prove of great importance. Suppose we have a single set of repeatable conditions $\mathfrak{S}_s$, then we can consider the set of conditions which are satisfied by a sequence of n repetitions of $\mathfrak{S}_s$ together with some new spacing condition s'. This we write $(\mathfrak{S}_s^n)_{s'}$. For example, suppose we are interested in problems about successive days of rain. We might take $\mathfrak{S} = \ldots$ is a day, i.e. a period between sunrise and sunset. An outcome of $\mathfrak{S}$ would be 'rain' or 'no rain'. Let $s =$ is separated by 24 hours and $s' =$ is separated by a year. Then a sequence of repetitions of $\mathfrak{S}_s$ would be a sequence of successive days and a sequence of repetitions of $(\mathfrak{S}_s^{28})_{s'}$ would be a sequence of corresponding periods of 28 days in successive years. Finally, if we write $\mathfrak{S}_s^n$ by itself we shall assume a spacing condition defined as follows. Consider a sequence of repetitions of $\mathfrak{S}_s$ and divide it into successive groups of n. This sequence of groups will be considered as a sequence of repetitions of $\mathfrak{S}_s^n$. Note that $\mathfrak{S}_s^n$ is a repeatable set of conditions. We can also define $\mathfrak{S}_{1s} \wedge \ldots \wedge \mathfrak{S}_{ns}$ in an analogous fashion.

If the above account is accepted, we can now definitely answer the question posed earlier of whether repeatability implies independence. The answer is 'no', as can be shown by constructing a set of conditions which are repeatable according to the above definition but whose outcomes are definitely not independent.

(ii) Repeatable conditions with dependent outcomes

We shall describe our examples by specifying the set of conditions $\mathfrak{S}$ and the spacing condition s.

(i) *Gas Example.* $\mathfrak{S} =$ take a fixed volume of gas at a fixed temperature. Consider a small cube in the centre of the gas at a fixed time. The outcome of $\mathfrak{S}$ is the number of molecules in this cube.

$s_t =$ repeat $\mathfrak{S}$ at time points $t_0, t_1, \ldots t_n, \ldots$ where $t_n > t_{n-1}$ and $t_n - t_{n-1} = t$ for some fixed t. We actually have here an infinite number of repeatable conditions depending on the value given to t. If t is sufficiently large the outcomes will be independent. This ceases to hold for small t, as the number of molecules in the cube at time t_n will then be only slightly different from the number at t_{n-1}. There is however a continuity between the two cases.

It will be useful sometimes to consider a more specific version of this. Let us suppose that the actual values of the quantities are so chosen that we usually have between 0 and 50 molecules in the cube, and so that the number of molecules in the cube changes by an average of about 3 from one time point to the next. The set Ω of all *a priori* possible results will be the set $(0, 1, \ldots, N)$ where N is some upper bound determined by the size of the molecules. Intuitively we can see that the probability of getting 7 is much higher when the previous result is, say, 4 or 3 than when it is 50.

(ii) *Game of Red or Blue.* $\mathfrak{S} =$ take a slate on which an integer $n(+\text{ve}, -\text{ve}, \text{or zero})$ is recorded. Toss a fair coin. Replace n by $n + 1$ if the result is heads and by $n - 1$ if it is tails. The result is the final number recorded.

$s =$ start with $n = 0$, and consider successive tosses of the coin. (Note that this spacing condition contains an initial condition and is of an abstract character.) There is sometimes a resistance to speaking of repetitions here because the 'repetitions' differ by the initial number n. However, we have already in effect met this objection. Two repetitions do not have to be the same in *all* ways but only in the specified set of ways.

We can imagine these conditions defining a game between two players, and the number n as recording the gains (or losses) of one of the players. Considering this interpretation, we can call any negative result a 'red', and any non-negative result a 'blue'. This explains the name 'Red or Blue'. A mathematical discussion of the game is given by Feller (1957, Ch. III) and the game is used as an example by Popper (1957a) to illustrate an argument against the subjective theory of probability. Once again the outcomes of this set of repeatable conditions are evidently dependent, for if we get a result $+100$, the result of the next repetition spaced according to s can only be $+99$ or $+101$. More generally it can be seen that any Markov chain will give us an example of the required type.

It could be objected against the red or blue example that the dependence on the number of previous goes is clear so that we shouldn't speak of repetitions. On the other hand consider the following example. An experimenter is given a black box with a dial on which numbers appear, and with a handle. He pulls the handle and observes the number which appears. This surely constitutes a well-defined repeatable experiment. The mechanism inside the box which produces the number is however a game of red or blue (as might in due course be discovered). This example may still seem rather artificial, but there is no reason why something very similar should not occur in practice. Imagine an atomic scientist counting at hourly intervals the number of particles produced in a minute by some set-up. Once again the underlying mechanism might be of the 'game of red or blue' variety, but the scientist would consider himself to be repeating a well-defined experimental procedure to get his readings.

It is now time to consider an objection to the above line of reasoning.[1] It runs as follows: 'Of course it is perfectly possible to define repeatability in some artificial way so that it does not entail independence. However, such a definition does not correspond at all to the way we normally use the notion of "repeatability". You have only succeeded in making a terminological distinction which has no interest or point.' My reply is this. My own intuitions as regards language tell me that the definition of repeatability I have given above is closer to our ordinary use of the term than a corresponding definition embodying independence would be. However, I am prepared to admit that ordinary usage may not be sufficiently clear to decide the point. Nonetheless it seems to me desirable to define repeatability as I did above because this definition enables us to ask the following question which has, I believe, a certain interest. Granted that we associate probabilities with the outcomes of certain sets of repeatable conditions, should we require that repetitions of these conditions be independent (that the conditions be repeatable and independent), or is repeatability alone sufficient? This is essentially a question about whether we should limit our application of the theory of probability in certain ways or not. The answer does not seem to me to be immediately obvious. I shall call the suggestion that we should require independence as well as repeatability the Axiom of Independent Repetitions, and will now argue that it should be adopted.

(iii) Arguments against the axiom of independent repetitions

Let us start, however, with some arguments against the axiom which we will try to rebut. It might be claimed that the axiom represents a needless truncation of the calculus of probability. For example, in his discussion of the game of red or blue, Popper

---

[1] I am indebted to Professor Popper for this objection.

makes the reasonable sounding observation that prob(red) = prob(blue) = $\frac{1}{2}$. However, if we adopt the axiom of independent repetitions, prob(red, $\mathfrak{S}_s$) = prob(blue, $\mathfrak{S}_s$) = $\frac{1}{2}$ (using our original notation) would not be admissible probabilities. Thus we are apparently ruling out harmless and useful probabilities.

Certainly we want to be able to say in some sense that prob(red) = prob(blue) = $\frac{1}{2}$, but we can do this without introducing probabilities relative to repeatable but dependent conditions. This can be shown quite easily using our concept of a spacing condition. Let $\mathfrak{S}_s$ be the original conditions of the game. s specifies that we consider a sequence of successive goes. Let s' specify that we consider a sequence consisting of every 1,000,000th go. Then the set of conditions $\mathfrak{S}_{s'}$ is independent as well as repeatable and we can introduce prob(blue) = prob(red) = $\frac{1}{2}$ relative to *these* conditions. Another way of proceeding would be to alter the conditions $\mathfrak{S}$ slightly to $\mathfrak{S}'$, say, so that they cover a number of different games of red or blue with different players and at different stages of play. We then take for spacing condition u the requirement that successive results are taken from different games. Once again $\mathfrak{S}'_u$ are independent as well as repeatable and we can take prob(blue) = prob(red) = $\frac{1}{2}$ relative to these conditions. Let us now turn to our example about the gas and the specific case introduced at the end. We might want here to introduce the probability of getting a result 7. This suggests that if we call the original repeatable but dependent conditions $\mathfrak{S}_s$ we should introduce prob(7) relative to them. However, in exactly analogous fashion to the 'red' or 'blue' case we can introduce this probability relative to repeatable and independent conditions. We can either consider every millionth result ($\mathfrak{S}_{s'}$) or a series of experiments on different (though similar) volumes of gas ($\mathfrak{S}'_u$). In this case there is, besides, a third way of proceeding. Let us add to our original conditions $\mathfrak{S}_s$ the additional condition: 'the previous result was R' where R can take any value between 0 and N. Let this set of conditions be denoted by $\mathfrak{S}_s \wedge R$. $\mathfrak{S}_s \wedge R$ is a repeatable and independent set of conditions for each R. We can thus introduce instead of the single prob(7), $N+1$ conditional probabilities for getting 7, namely $p(7, \mathfrak{S}_s \wedge 0), \ldots$ $p(7, \mathfrak{S}_s \wedge N)$; or as they could be more simply denoted: $p(7,0) \ldots$ $p(7,N)$.

We can deal in the same way with a similar objection. 'You say that we should confine ourselves to sequences of independent repetitions. But the mathematical theory of probability does not deal only with the case of independence. It also considers sequences of dependent events (Markov chains for example), and obtains many interesting results. Your suggestion is tantamount to ruling out this interesting field.' Once again, however, by suitably altering the underlying conditions we can get what we need. Suppose, for example, we want to consider the probabilistic relations involved in a game of red or blue of N goes for some large N (i.e. to consider a sequence of N repetitions of $\mathfrak{S}_s$). We introduce a new set of conditions $(\mathfrak{S}_s^N)_v$ as follows. Let $\mathfrak{S}_s^N$ as usual be satisfied by an n-tuple of events $(e_1,...,e_N)$ each of which satisfies $\mathfrak{S}_s$ and which are spaced according to s. Let v (the spacing condition) specify that we start each repetition in the zero position. Then relative to $(\mathfrak{S}_s^N)_v$ (which are in-dependent as well as repeatable) we can introduce all the probabilities we need for the required study. This is in effect what Feller does in his account of the matter (cf. 1957, p. 73), though, as he does not mention the underlying conditions at all, the point is not made explicitly. Granted then that by suitable manipulation of the repeatable conditions we *can* ensure that the postulate of independent repetitions is satisfied it can still be asked whether we *need* to do so. Is there any point in such a procedure? I claim that there is and will now give my arguments.

(iv) Main argument for the axiom of independent repetitions
Suppose that contrary to the postulate of independent repeti-tions we introduce the probabilities prob (red, $\mathfrak{S}_s$) = prob (blue, $\mathfrak{S}_s$) = $\frac{1}{2}$ in the game red or blue. Let us repeat $\mathfrak{S}_s$ a large number N of times and suppose that 'red' occurs $M(\text{red})$ times. As usual we consider the relationship between $M(\text{red})/N$ and prob (red, $\mathfrak{S}_s$) (i.e. $\frac{1}{2}$). Now in this case (as explained above), by introducing the spacing condition v we have a set of conditions $(\mathfrak{S}_s^N)_v$ for which the postulate of independent repetitions is satisfied. A particular value of $M(\text{red})/N$ could be considered as a result of these conditions. Thus we can quite legitimately consider the probabilities of events such as:

$$|M(\text{red})/N - \tfrac{1}{2}| > \epsilon \quad \text{where} \quad \epsilon > 0.$$

In fact these probabilities are determined by the underlying conditions defining the game red or blue. Calculations (cf. Feller, 1957, pp. 77–8) give the following results. Suppose we toss the coin once a second for a year, i.e.

$$N = 60 \times 60 \times 24 \times 365 = 31{,}536{,}000.$$

Let $p = \text{prob}(|M(\text{red})/N - \tfrac{1}{2}| > \epsilon)$. Then we have ($\epsilon$ correct to 3 decimal places)

p	ϵ
0·9	0·078
0·8	0·155
0·7	0·227
0·6	0·294
0·5	0·354
0·4	0·405
0·3	0·456
0·2	0·476
0·1	0·494

These results are I think very striking. Even with this colossal number of repetitions (over 30 million) there is a probability of 0·9 that the observed relative frequency, if evaluated to the first decimal place, will differ from the 'probability' 0·5. There is a probability of over 60 per cent that the difference will be more than 50 per cent of the 'probability'. Finally there is a probability of over 30 per cent that the relative frequency evaluated to 1 decimal place will give a value (either 0 or 1) as different as possible from the 'probability'.

Let us compare these results with equation (1) of Chapter 4 (p. 80). This stated that for large n, $m(A)/n \doteq p(A)$, or rather that this result must hold with large probability, i.e. be nearly certain. It was shown then that both Kolmogorov and Cramér held this equation to be true, and I remarked that any probability theory in which the equation was not valid would be extraordinary and most unacceptable. However, if we give up the postulate of independent repetitions in some cases at least, equation (1) is abrogated in a very drastic manner. This is an 'intuitive contradiction', and I conclude that we ought to accept the disputed axiom. Such is my main argument for the axiom of independent repetitions. I will, however, give a second argument concerned with the probabilities of single events in Chapter 7. Our next step will be to set up the axioms of probability in such a manner that they include a formalized version of the axiom of independent repetitions. We will then formulate (roughly) a methodological rule which I claim must be used whenever probability statements are tested by frequency evidence. From the axioms of probability we will, using this methodological rule, derive the law of stability of statistical frequencies.

(v) Derivation of the law of stability of statistical frequencies

First some notation. According to my account probability theory deals with *probability systems*. These are ordered quadruples $(\mathfrak{S}_s, \Omega, \mathfrak{F}, p)$ where $\mathfrak{S}_s$ is a set of repeatable conditions with spacing parameter s (as defined above), Ω is the set of possible outcomes of $\mathfrak{S}_s$, $\mathfrak{F}$ is a σ-field of subsets of Ω including Ω itself, and p is a set function defined on $\mathfrak{F}$. Further, let Ω^n denote the set of n-tuples $(\omega_1, \ldots, \omega_n)$ where each $\omega_i \in \Omega$. Let $\mathfrak{F}^n$ be a σ-field of subsets of Ω^n defined as follows: we consider the set (S say) of all Cartesian products $A_1 \times \ldots \times A_n$ where each $A_i \in \mathfrak{F}$ and take $\mathfrak{F}^n$ to be the minimum σ-field containing S. We can now state the axioms of probability theory as follows:

Axiom I (Kolmogorov's axioms). p is a non-neg, σ-additive set function on $\mathfrak{F}$ s.t. $p(\Omega) = 1$.

Axiom II (axiom of independent repetitions). If $(\mathfrak{S}_s, \Omega, \mathfrak{F}, p)$ is a probability system, so also are the quadruples $(\mathfrak{S}_s^n, \Omega^n, \mathfrak{F}^n, p_n)$ for all $n > 1$, where the measure p_n on $\mathfrak{F}^n$ is the n-fold product measure of the measures p on $\mathfrak{F}$.

Axiom II corresponds to von Mises' axiom of randomness in the same loose way that von Mises' axiom of convergence corresponds to the standard Kolmogorov axioms. I noted as point (3) (see p. 88) in my table of differences that there was nothing corresponding to the axiom of randomness in Kolmogorov's system. This missing item has now been supplied. It also becomes appropriate to repeat in this new context a remark made by von Mises in his original paper (1919, p. 70):

Something similar occurs in the case of certain problems investigated by Borel and others (for example problems about the appearance of particular integers in the infinite decimal

8

expansion of irrational numbers), where the question of the satisfaction or non-satisfaction of Postulate II is senseless.

In the applications of probability theory to number theory it is a question of the application of one pure mathematical concept (the probability space) in the investigation of others. Remarks about experimental conditions can only be metaphorical in this context, and so the notions of probability system and the axiom of independent repetitions are inappropriate here. Actually another concept of pure mathematics – that of the conditional probability space, introduced by Rényi – is more appropriate than that of the probability space in some of these contexts.

We cannot derive the results we want about relative frequencies from these axioms alone. We need some further rule to enable us to carry out this task, though it is not difficult to see what this rule must be. The point is that in order to test statistical hypotheses using frequency evidence it is necessary to neglect certain small probabilities in certain circumstances; and this is indeed just what statisticians do. Consider the χ^2-test for example. Suppose a 5 per cent level of significance is adopted. It is then calculated that if the hypothesis is true, the χ^2-statistic will lie with 95 per cent probability in a certain interval – (a,b) say. The hypothesis is regarded as corroborated if $\chi^2 \in (a,b)$ and falsified if $\chi^2 \notin (a,b)$. We are in effect *predicting* that $\chi^2 \in (a,b)$ and checking this prediction against the facts. Conversely, if we wanted to *explain* some frequency phenomenon summed up in the statement $\chi^2 \in (a,b)$ we could use the statistical hypothesis in question. More generally we can say that if we neglect this probability of 5 per cent we can in a certain sense 'deduce' that $\chi^2 \in (a,b)$ and this deduction can be used to explain or predict in accordance with the deductive model. What it comes down to is this. When the calculus of probabilities is used to explain certain statements about frequencies, a methodological rule is adopted to the effect that one can neglect certain small probabilities in certain circumstances. This situation sets, I believe, two tasks for the philosopher: (1) to formulate explicitly this rule which underlies statistical practice and (2) to consider what justification (if any) can be offered for the rule.

These problems will be studied in detail in Part III, Chapters

9–11. For the moment I shall content myself with drawing attention to the following difficulty. We cannot formulate the rule as being that of 'neglecting small probabilities', because this would lead to absurd results in some cases. Suppose we toss a coin a million times and observe exactly the same number of heads and tails. We would certainly regard this result as corroborating the hypothesis that prob (heads) $= \frac{1}{2}$. On the other hand the probability of this result on the given hypothesis is only 0·0008 (to 1 sig. fig.). We cannot here neglect a small probability of less than a thousandth, and yet in a statistical test we can (quite validly) neglect a probability of one-twentieth (5 per cent significance level). This is why the methodological rule must be of the form: one can neglect such and such small probabilities in *such and such circumstances*. For the moment we will assume that we have made a correct application of this putative rule if what we do is close enough to statistical practice. This will enable us to apply the above considerations to the example in hand.

Consider therefore an arbitrary probability system $(\mathfrak{S}_s, \Omega, \mathfrak{F}, p)$ and let $A \in \mathfrak{F}$ with $p(A) = p$. Repeat $\mathfrak{S}_s$ n times and suppose A occurs $m(A)$ times, then from our two axioms we obtain at once

$$\text{prob}\,(m(A) = r) = {}^nC_r\,p^r(1-p)^{n-r}.$$

By the central limit theorem it now follows that for large n, $m(A)/n$ is approximately normally distributed with mean p and standard deviation $(p(1-p)/n)^{1/2}$. The normal approximation is only introduced for computational convenience. More sophisticated and accurate approximations are of course available (see, for example, Uspensky, 1937, Chapter VII, pp. 119–34), but the present simple method will, I think, suffice for our purposes. If we now agree to neglect a probability π in accordance with our methodological rule we obtain that

$$\left| m(A)/n - p \right| < \lambda_\pi (p(1-p)/n)^{1/2} \qquad (1)$$

where λ_π is a constant depending on π; in fact the π percentage point of the normal distribution.

This is the law of stability of statistical frequencies, or rather it is the law in a more precise form because (1) does not merely tell us that $m(A)/n$ is near p for large n, it gives us an estimation of how close to $p\, m(A)/n$ will be for specified values of n. Thus in deducing the law of stability of statistical frequencies we have rendered it more precise.

To this it will be objected that the value of λ_π is still vague so that we have hardly made the law more exact. But this is not quite fair. In practical tests statisticians almost invariably neglect probabilities of 1, 5, or 10 per cent. For the normal distribution we have correspondingly $\lambda_1 = 2\cdot576$, $\lambda_5 = 1\cdot960$, $\lambda_{10} = 1\cdot645$. To see how much variation this introduces let us take a particular example. Suppose $p = 0\cdot5$ and we are interested in knowing how large n must be so that $m(A)/n = p$ correct first to 1 decimal place, and then to 2 decimal places, i.e. so that $|m(A)/n - p| < 0\cdot05$ resp. $0\cdot005$. A simple calculation shows that we must have $n > \lambda_\pi \times 10^2$ resp. $\lambda_\pi \times 10^4$. The three usual values of λ_π give 663, 382, 272 for 1 decimal place [and 66,300, 38,200, 27,200 for 2 decimal places. We conclude that $m(A)/n$ will be equal to $0\cdot5$ correct to 1 decimal place] after between approximately 250 and 650 repetitions, and equal to $0\cdot5$ correct to 2 decimal places after between approximately 25,000 and 65,000. These results are not completely exact but they are a good deal more exact than saying merely that $m(A)/n$ will be approximately equal to $0\cdot5$ for large n. We also conclude that $m(A)/n$ will converge to p at the rate at which $1/(n)^{1/2}$ converges to 0. This again is much more accurate than merely saying that $m(A)/n$ will converge to p. As a matter of fact I believe that we can fix the value of λ_π more precisely than is usually done in statistical practice. However, to justify this assertion it would be necessary to give a full discussion of the methodological rule concerned with neglecting small probabilities. As I said earlier this discussion will be postponed to Part III.

It is interesting to quote again in this context some remarks made by von Mises when describing the law of stability of statistical frequencies. He says (1928, p. 14):

> If the relative frequency of heads is calculated accurately to the first decimal place, it would not be difficult to attain constancy in this first approximation. In fact perhaps after some 500 games, this first approximation will reach the value of $0\cdot5$ and will not change afterwards. It will take much longer to arrive at a constant value for the second approximation,

calculated to two decimal places.... Perhaps more than 10,000
casts will be necessary to show how the second figure also
ceases to change, and remains equal to 0, so that the relative
frequency remains constantly 0·50.

It will be seen that our theoretical predictions agree with what
von Mises says except that his 10,000 casts is an underestimate.
According to our calculations we would need at least 25,000 casts
to ensure that $m(A)/n = 0·50$ correct to 2 decimal places. These
remarks of von Mises could, if he had generalized them, have
led to a more precise statement of the law of stability of statistical
frequencies, but he did not in fact produce such a statement.
Some mathematical work was necessary before the statement of
a purely empirical result could be reached.

(vi) Comparison with Newtonian mechanics

Let us next compare the situation here with the paradigm case
of Newtonian mechanics. It must be admitted that the parallel
is remarkably close. In the case of Newtonian mechanics we
have a system of axioms involving the new concepts of 'force'
and 'mass'. This system was tested out by deriving from it
certain well-known laws (Kepler's and Galileo's laws) which
did *not* involve 'force' and 'mass'. The derivation was accom-
plished by neglecting one mass in comparison with another. In
the case of probability theory as expounded above we have a
system of axioms involving the new concepts 'probability' and
'independence'. This system is tested out by deriving from it
the well-known 'law of stability of statistical frequencies' which
does not involve the new concepts. The derivation is accom-
plished by neglecting one probability in comparison with another.
So far there is an exact correspondence.

It is now time to point out a couple of differences. The first one
is not very important. Newtonian mechanics showed itself to
be a deeper theory by correcting Kepler's laws while explaining
them. Probability theory does not correct the law of stability
of statistical frequencies (which was anyway a very vaguely
formulated law), but does render it more precise while explaining
it. However, as I remarked above, 'rendering more precise' seems
just as adequate a criterion for depth as correcting the law
explained. The second difference is a more genuine one. The

approximation, mass(planet) + mass(sun) = constant, used in Newtonian mechanics was based on the physical assumption that mass(sun) ≫ mass(planet). Now it was perhaps necessary to make this assumption in order to test Newtonian theory initially. However, at a later stage it could be replaced by a different and more precise assumption which would lead to a more accurate approximation. The situation in probability theory is rather different. To begin with the decision to neglect a certain probability is not based on a physical assumption. From the theory itself we can calculate the size of the probability we are neglecting (say 5 per cent). Further, there is little chance of this approximation being replaced by a better one. The reason is simple. Probability statements are always (or nearly always) tested by frequency evidence. To carry out such a test we must infer from the probability statement a statement about frequencies and then compare this prediction with the actual frequencies observed. In order to make the inference mentioned here we must agree to neglect a small probability. Thus whenever we test out a probability statement, or use a probability theory to explain a frequency phenomenon, we have to neglect small probabilities, and the small probabilities neglected are always of the same order (about 5 per cent). In effect, whenever the probability calculus is applied we have to bind ourselves by a methodological rule telling us to neglect such and such small probabilities in such and such circumstances. This rule must be adopted since otherwise probability statements would not be testable (falsifiable), and we could not use them to obtain genuine scientific explanations. In mechanics the situation is quite different. We do not have to adopt a rule telling us to neglect one mass in comparison with another whenever we apply the theory.

(vii) Criticism of the views of von Mises and Kolmogorov on the relations between probability and frequency

The deduction of the law of stability of statistical frequencies of course establishes the connection between probability and frequency. Having given my own account of this connection, I next want to criticize the three alternative views mentioned in Chapter 4. My view differs from the 'frequency' theories (in the wide sense defined above) of von Mises and Kolmogorov because probability is identified with frequency neither in the formalism nor in any informal supplement. The link with experience is provided by neglecting small probabilities in certain circumstances; and we do not need an elaborate definition of probability in terms of frequency to justify this procedure. It is also interesting to examine in this context what Kolmogorov (1933, p. 4) calls: 'The Empirical Deduction of the Axioms'. Kolmogorov recognizes that probability theory is an empirical science. It is therefore necessary for him to give some evidence for his axioms. This he does as follows. Probabilities are (at least approximately) identified with relative frequencies m/n for large n. Thus if certain relative frequencies have a particular property, it is reasonable to postulate that the corresponding probabilities have the same property. In this way we can derive the axioms of probability from corresponding properties of relative frequencies. For example, Kolmogorov derives the additivity of probability thus (1933, p. 4):

> If, finally, A and B are non-intersecting (incompatible), then $m = m_1 + m_2$ where m, m_1, m_2 are respectively the number of experiments in which the events $A + B$, A, and B occur. From this it follows that
>
> $$m/n = m_1/n + m_2/n.$$
>
> It therefore seems appropriate to postulate that $p(A + B) = p(A) + p(B)$ (Axiom V).

A similar empirical deduction of the axioms is given by Cramér (1945, pp. 149 and 153).

We see here once again the influence of operationalist philosophy. The concepts of a theory (like probability) are supposed to be abstractions of certain observable quantities (in this case relative frequencies m/n for large n). The observable quantities obey certain empirical laws, and the postulates of the theory are the corresponding abstract versions of these laws. Kolmogorov's theory of probability differs mathematically from von Mises', but it is justified using the same philosophical presuppositions. As against these, our own approach is hypothetico-deductive. The axioms are not 'deduced' from any empirical results. They are postulated quite freely and then tested by deducing certain results about frequencies.

(viii) An answer to some arguments of Doob's

Let us now consider Doob's view, i.e. that the relation between probability and frequency is established by the laws of large numbers. This is in certain respects more similar to my own approach than the frequency views just considered, but it is still sufficiently different to be unacceptable. I have attributed this view to Doob, but I think it is held by many other mathematicians. For example, Rényi in the paper dealing with his new concept of a conditional probability space says (1955, p. 326):

> Conditional probability is in the same relation to conditional relative frequency as ordinary probability to ordinary relative frequency. This relation, which is well known as an empirical fact from every day experience, is described mathematically by the laws of large numbers.

Before criticizing this view, I would like to counter certain arguments of Doob. These were originally directed against von Mises' theory, but they in fact apply equally to my own approach.

Doob says (1941, p. 211): 'There can be no question of the need for any axiomatic development beyond that necessary for measure theory....' However, we have added an extra axiom (the axiom of independent repetitions) which is not an axiom of measure theory but which is designed to tie probability theory in with experience. He says further (1941, p. 217):

> The frequency theory reduces everything to the study of sequences of mutually independent chance variables, having a common distribution.... This point of view is extremely narrow. Many problems of probability, say those involved in time series, can only be reduced in a most artificial way to the study of a sequence of mutually independent chance variables, and the actual study is not helped by this reduction, which is merely a tour de force.

Now we have postulated that probabilities should only be introduced within probability systems $(\mathfrak{S}_s, \Omega, \mathfrak{F}, p)$. In most practical cases we can introduce a random variable ξ defined on Ω and replace consideration of the probability space $(\Omega, \mathfrak{F}, p)$ by consideration of this random variable and its distribution function, F say. However, by the axiom of independent repetitions, if we repeat $\mathfrak{S}_s$ we obtain a sequence of independent random variables with the same distribution F. Thus in a certain sense at least we have reduced 'everything to the study of sequences of mutually independent chance variables, having a common distribution'. However, we do not regard this point of view as being 'extremely narrow'. In order to counter Doob's view let us examine the chain of reasoning which led him to his position.

Suppose we are trying to formulate the problem of throwing a die within the language of the probability space $(\Omega, \mathfrak{F}, p)$. We could do it like this. Take Ω to be the set of sequences $(x_1, \ldots, x_n, \ldots)$ where x_n is an integer between 1 and 6 inclusive. x_1 represents the result of the first throw, etc. Now introduce random variables ξ_n defined on Ω by

$$\xi_n(\omega) = x_n \quad \text{where} \quad \omega = (x_1, \ldots, x_n, \ldots).$$

We postulate that these ξ_n are independent random variables having the same distribution defined by $p(\xi_n = i) = 1/6$ for $1 \leqslant i \leqslant 6$. We have now specified the problem mathematically and can proceed to calculate the probabilities required. Let us next consider a problem which involves dependence. We will take the game of red or blue again, though any example of a Markov chain, or of the time series mentioned by Doob, would do just as well. Here we would, as before, choose for Ω a set of sequences $(x_1, \ldots, x_n, \ldots)$. This time, however, x_n can take on any integral value (+ve, −ve or 0). x_1 represents the result of the first go of the game, etc. Now introduce random variables ξ_n defined on Ω by

$$\xi_n(\omega) = x_n \quad \text{where} \quad \omega = (x_1, \ldots, x_n, \ldots).$$

This time we postulate that the ξ_n are random variables with different distributions and also that certain dependence relations hold between them. From these assumptions we can go on to calculate the properties of the game in which we are interested. It now looks as if the cases of independent random variables and of dependent random variables are essentially similar, and that independence is no more fundamental than dependence.

To this argument I reply as follows. Consider our sample space Ω in the second case (the game of red or blue). What meaning does Ω have? I claim that we must as usual take Ω

to be the set of *a priori* possible outcomes of a set of repeatable conditions $\mathfrak{S}$. In this case $\mathfrak{S}$ is the set of conditions defining a complete (infinite) game of red or blue. Repetitions of $\mathfrak{S}$ would be separate infinite games of red or blue all starting from the zero position. Now there may be a different way of interpreting Ω here, but none has as far as I know been suggested, and I cannot devise one myself.

So then Ω is the set of outcomes of repeatable conditions $\mathfrak{S}$. But once again we can ask what is meant by 'repeatable' here and whether 'repeatability implies independence'. If our earlier argument is correct, we conclude that repeatability does not imply independence, but that we have to postulate independence as well as repeatability. Thus the analysis of dependent events has to be based on the consideration of certain independent events. In the case of the game of red or blue, these independent events are independent games starting at the same zero position and continuing indefinitely. More generally, authors who write on stochastic processes involving dependent probabilities very often speak of 'independent realizations of the process' (an implicit use of the axiom of independent repetitions, I would claim). Correspondingly they distinguish 'averages over the process itself' and 'averages over independent realizations of the process'. This is a nice illustration of the way in which considerations of independence underlie the analysis of dependence.

(ix) Criticism of the 'laws of large numbers' view

I will now criticize the view that the relation between probability and frequency is established by the laws of large numbers (the 'laws of large numbers' view as we shall call it). Let us consider a relative frequency m/n and a probability p, and suppose we can prove a strong law of large numbers of the form:

$$\text{prob}\left(\lim_{n\to\infty} m/n = p\right) = 1. \qquad (2)$$

Does this establish the connection between probability and frequency? In order to infer from (2) that

$$p = \lim_{n\to\infty} m/n \qquad (3)$$

we have to employ a rule of the form: 'Regard as certain a probability of 1' or equivalently: 'neglect zero probabilities'. This illustrates that if we take a non-frequency view, i.e. we do not regard probability as in any sense defined in terms of frequency, then in order to establish the connection between the two we have to adopt some methodological rule telling us that we can neglect such and such probabilities in such and such circumstances when applying the calculus. It may look as if the rule implicitly used here (viz. neglect zero probabilities) is much better than our own rule which involves neglecting probabilities of 5 per cent or so. However, this argument can be turned on its head. The methodological rule must be such as to apply to all the uses of probability in statistical inference. Now if we adopted the rule 'neglect only zero probabilities', statistical inference could never begin. In statistics we have to be more liberal and agree to neglect probabilities of about 5 per cent, and therefore, I claim, we should adopt the *same* rule when dealing with the link between probability and frequency in the foundations of the subject. Thus I reject the 'law of large numbers' view because it makes implicit use of a methodological rule different from that employed in most statistical applications.[1]

My second objection is that the relation inferred is the limiting frequency result (3). We can therefore raise all the customary objections against the use of limits here: namely that in practice there are only a finite number of trials and more sophisticated versions of this argument. This might be countered by a *'tu quoque'*. After all, in my own derivation I replace a binomial distribution by the normal curve to which it tends in the limit. Now of course I do not want to ban the use of limits everywhere in applied mathematics, but I would claim that some uses of limits are much more suspect than others. In particular I would maintain that replacing a binomial distribution for large n by a normal curve is all right, but that the use of limits as in equation (3) is much more suspect. The former use of limits is a mathematical approximation introduced for computational purposes. It is a purely mathematical matter whether, and to what extent, the binomial distribution for large n approximates to the normal

---

[1] This argument really needs to be buttressed by an explicit formulation of the methodological rule underlying statistical practice. Admittedly we have not done this yet, but see Part III, Chapters 9–11.

curve. We can estimate the degree of the approximation mathematically. Better still we could dispense with it altogether if we were working out results on a computer.

When we apply equation (3), however, we are not making a purely mathematical approximation. We also need to make an empirical or physical assumption to the effect that a large number of repetitions, say 500, is adequately represented by an infinite sequence. I fully agree that such assumptions do have to be made very often, but I would argue that where they can be avoided, they should be avoided. This is the case here as my earlier discussion showed.

The dangers involved in making assumptions to the effect that the 'large' finite can be approximated by the infinite are well illustrated by von Mises in his 1963 volume. He claims there that all applications of probability theory depend on the silent assumption of rapid convergence:

> Our silent assumption is that...*in certain known fields of application of probability theory* (games of chance, physics, biology, insurance etc.) *the frequency limits are approached comparatively rapidly.* (p. 108)

If we apply probability theory to some new area such as the study of supposed parapsychological phenomena, the results it yields will be untrustworthy because the silent assumption may not be satisfied here. 'However, in "new" domains where we do not know whether "rapid convergence" prevails, "significant" results in the usual sense may not indicate any "reality". It seems, for example, a wrong approach to try to "prove" parapsychological phenomena by statistical methods' (1963, p. 110). But are we right to reject the results obtained by statistical methods in a certain area on the grounds that some 'silent assumption' may not hold? And how anyway can we tell whether the silent assumption is satisfied or not?

In our approach this 'silent' assumption is replaced by entirely audible assumptions from which rapid convergence is actually deduced. We assume that the probabilities are defined on independent and repeatable conditions, that they constitute a measure and that we can neglect small probabilities in certain circumstances. Our deduction of the law of the stability of statistical frequencies then establishes rapid convergence. Of course in any particular case any one of these assumptions could fail (except perhaps the assumption that small probabilities can be neglected, which is normative in character); and indeed rapid convergence *can fail* as we saw in the game 'red or blue'.

The case of parapsychological phenomena presents no special problems. Suppose someone is guessing cards and we want to investigate whether he is using telepathy. As usual we would set up a null-hypothesis that his successful guesses are produced by some chance mechanism: say there is a probability p of him guessing right on each trial, where p = relative frequency of a given type of card in the pack, and the trials are independent. We then test out this hypothesis using the frequency evidence. If it is falsified this does not of course *prove* the existence of telepathy. It merely shows that the original chance hypothesis is false. The deviations may be due to telepathy, to some more natural cause, or it may be that a chance mechanism does hold – only a more complicated one than that which we first postulated. The whole situation is exactly the same as when statistics are applied to, say, agricultural experiments about wheat yields.

My third objection to the law of large numbers view is really just a particular instance of the second. It is that we can often prove a law of large numbers where the kind of rapid convergence of frequency to probability which we require in practice does not occur. Once again the game of red or blue will furnish an example of this. We will modify it slightly by putting on two 'reflecting barriers'. Let us suppose that when $n = 40$ million and we get 'heads' we retain n instead of going to $n + 1$, whereas if we get 'tails' we go to $n - 1$. Similarly we arrange not to go below -40 million. These reflecting barriers are sufficiently widely spaced for the paradoxical results described above to hold still. Thus $m(\text{red})/n$ does not converge 'rapidly' or 'in the practical sense' to $\frac{1}{2}$ (i.e. prob (red)) as n tends to ∞. However, $m(\text{red})/n$ does converge to $\frac{1}{2}$ with probability 1, i.e. a strong law of large numbers holds. This last result is, I think, intuitively obvious but we can deduce it from a general theorem. Doob (1953, p. 219) gives the following result. If $\xi_1, \ldots, \xi_n, \ldots$ are random variables constituting a Markov chain and f is a Baire function, then under Hypothesis (D) and if there is only one ergodic set,

$$\lim_{n \to \infty} 1/n \sum_{m=1}^{n} f(\xi_m) = E(f(\xi_1))$$

with probability 1. As for hypothesis (D), Doob remarks (1953, p. 192) that it is always satisfied if the ξ_i can only take on a finite number of values (as here). We also have only one ergodic set in Doob's sense. Now define $f(x)$ as follows:

$$f(x) = 1 \quad \text{if} \quad x < 0$$
$$\quad\quad = 0 \quad \text{if} \quad x \geqslant 0.$$

Then $\sum_{m=1}^{n} f(\xi_m) = m(\text{red})$ and we deduce $\lim_{n \to \infty} m(\text{red})/n = \frac{1}{2}$ with probability 1 as required.

(x) The significance of the laws of large numbers

If the above thesis is correct and the laws of large numbers do *not* establish the connection between probability and frequency, the question arises: 'what *is* the significance of these laws?' I answer that the laws have no importance for probability theory considered as an empirical science. Their merit lies only in their mathematical elegance. They are thus only significant for probability theory qua pure mathematics.

This view can be contrasted with that of Gnedenko and Kolmogorov who write (1949, p. 1): '...the epistemological value of the theory of probability is revealed only by limit theorems'. This is also an appropriate point to discuss some further interesting remarks of Kolmogorov. He says (1933, p. 8):

> Indeed, as we have already seen, the theory of probability can be regarded from the mathematical point of view as a special application of the general theory of additive set functions. One naturally asks, how did it happen that the theory of probability developed into a large individual science possessing its own methods?

To this question he answers (p. 8): 'Historically, the independence of experiments and random variables represents the very mathematical concept that has given the theory of probability its peculiar stamp.' In a sense this question is a little misformulated. Probability theory had developed as a special science dealing with particular phenomena (games of chance and the rest) for at least 250 years before the invention of measure theory. And by the time additive set functions were introduced for the first time probability theory was already a 'large individual science'. However we can, I think, put Kolmogorov's point like this. Measure theory and the theory of additive set functions developed as pieces of pure mathematics which were off-shoots of classical analysis. What ideas from the empirical science of probability contributed to the development of this branch of pure mathematics?

Among others we may distinguish two important ideas which are related to the preceding discussion. First we have (as Kolmogorov says) the notion of independence, and second the idea of proving general theorems of the law of large numbers type. We will now elaborate these points. A large section of classical analysis deals with summing infinite series of real numbers $\sum_{n=1}^{\infty} a_n$. Now an obvious generalization of this is to consider summing series of functions $\sum_{n=1}^{\infty} f_n$ which are all defined on some abstract space Ω. If we defined convergence for the partial sums $\sum_{n=1}^{N} f_n$, as convergence of $\sum_{n=1}^{\infty} f_n(\omega)$ for each $\omega \in \Omega$, then no real generalization of the ideas of classical analysis would be achieved. If, however, we assume that a measure is defined on Ω and that the f_n are measurable functions, we can then consider for example the convergence of $\sum_{n=1}^{N} f_n(\omega)$ for all $\omega \in \Omega$ except for ω in a set of measure zero. In this way a new notion of an infinite sum of functions is reached. In a similar way we can imagine the notions of convergence in measure and in mean developing. However, before we can prove interesting results about convergence of sequences of functions f_n, we have to impose some extra conditions on the functions. What conditions should these be? It is here that probability theory was able to suggest a new move.

One of the basic problems of probability was to investigate the relations between the frequency $m(A)/n$ of an event A and its probability $p(A)$. Let us introduce the random variable $f_i = 1$ if A occurs on the ith repetition and $= 0$ if A does not occur. Then $m(A)/n = (f_1 + \cdots + f_n)/n$ and the f_i are independent random variables. Analogy with this situation suggests that in functional analysis we should look for limit theorems of the form $\lim_{n \to \infty}(f_1 + \cdots + f_n)/n$ where the f_i are independent functions. Such investigations were carried out systematically in the period 1924–33 by Kolmogorov, Khintchine and Lévy, and later by other authors. As a result there is now an elegant theory of convergence of series and sequences of independent random variables which constitutes an interesting extension of the classical theory of convergence of series and sequences of real numbers.

The next point I want to make is that these ideas, although in fact taken from probability theory, could have developed spontaneously within functional analysis itself. After all, the probabilistic notion of independence corresponds to the notion of product measure, and this could certainly have been (and indeed was) evolved without using probabilistic ideas. Finally, the idea of considering $(f_1 + \cdots + f_n)/n$ could well have been suggested by the notion of a Cesaro sum. This case shows how pure mathematics can sometimes develop as a self-contained system, but sometimes makes use of ideas drawn from the sciences.

Deduction of the Law of Excluded Gambling Systems: The Role of Randomness in Probability Theory

(i) Deduction of the law of excluded gambling systems

We now come to the question of deducing the law of excluded gambling systems. Roughly this law may be stated as follows: Suppose we have a probability system $(\mathfrak{S}_s, \Omega, \mathfrak{F}, p)$ and suppose $A \in \mathfrak{F}$ with $p(A) = p$. Event A has probability p in a sequence of repetitions of $\mathfrak{S}_s$ and the law states that we cannot increase our chance of getting A by selecting a subsequence from the sequence of repetitions. To get a more precise statement let us introduce a sequence of random variables $\xi_1, \xi_2, \ldots, \xi_n, \ldots$ where $\xi_n = 1$ if A occurs at the nth repetition, and $= 0$ otherwise. By the axiom of independent repetitions these random variables are independent and have the same distribution, namely $p(\xi_n = 1) = p$, $p(\xi_n = 0) = 1 - p$ where $0 \leqslant p \leqslant 1$. A gambling system (using Church's approach) is an effective algorithm enabling us to calculate a value $g_i = 1$ or 0 given i and the results of $\xi_1, \ldots, \xi_{i-1}$. If $g_i = 1$ we bet that the ith trial will give A. If $g_i = 0$, we do not bet. Suppose we employ the gambling system on a finite initial segment n of our sequence of repetitions. Let the set $I_n = \hat{i}$ $(g = 1$ and $i \leqslant n)$ contain n' members. Let the number of ξ_i where $i \in I_n$ and which give an observed value of 1 be m'. Then if $m'/n' > p$ the gambling system will have been successful for this initial segment. A gambling system can be said to be successful in general if for all or at least most large n' it is successful.

Why can no gambling system be successful? Consider the set (ξ_i) s.t. $i \in I_n$. This set consists of n' independent random variables with the identical distribution $p(1) = p$ and $p(0) = 1 - p$.

Further n' is large. Therefore just as in Chapter 5 above we can deduce that the observed frequency ratio m'/n' lies in a narrow interval round p; so to obtain $m'/n' > p$ for all or even most large n' is impossible. This is the law of excluded gambling systems.

Once again our finitistic approach contrasts with Doob's infinite method. Doob dealt with the problem of excluded gambling systems in his 1936 paper 'A Note on Probability'. His results are summarized on pp. 144–7 of his book on *Stochastic Processes* (1953). Doob considers a sequence of random variables $\xi_1, \ldots, \xi_n, \ldots$ which are independent and identically distributed. He obtains a subsequence $\xi'_1, \ldots, \xi'_n, \ldots$ by a selection procedure which we will not describe. He then thinks that the law of excluded gambling systems is contained in the statement that the random variables ξ'_1, $\xi'_2, \ldots$ have the same probability properties as $\xi_1, \xi_2, \ldots$ so that in particular

$$\lim_{n \to \infty} (\xi'_1 + \cdots + \xi'_n)/n = E(\xi'_1) = E(\xi_1) \qquad (1)$$

with probability 1. We can criticize this approach in exactly the same way as the law of large numbers view discussed in Chapter 5. First of all we should avoid approximating the finite by the infinite wherever possible, and we can do so here. Secondly, to obtain practical results from (1), we have to use a methodological rule for neglecting probabilities which does not correspond to the rule normally applied in statistics.

(ii) Independence and gambling systems

Our deduction of the law of excluded gambling systems depends crucially on the assumption of independence for the original sequence of random variables: $\xi_1, \xi_2, \ldots, \xi_n, \ldots$. Conversely a gambling system can be considered as a test of independence. To see this let us consider again the elementary test described in Chapter 5. The test consisted of observing the number m of 1's in the first n observed values of the sequence $\xi_1, \ldots, \xi_n, \ldots$ of random variables. If m/n differed too much from p, our probability hypothesis involving independence and identical distribution was regarded as falsified. It is easy, however, to imagine a finite sequence of 0's and 1's which would pass this test, but whose form would lead us to conjecture that the hypothesis was false. Consider for example the case $p = \frac{1}{2}$. If we obtained the sequence 0101...01 in 5,000 tosses, the hypothesis would certainly pass our elementary test but it would be obvious that such a result could not have been given by a sequence of *independent* random variables. The dependence of one result on its predecessor is quite clear. This observation naturally suggests a second test which would have succeeded in falsifying the hypothesis. Let us now consider not the whole sample but a subsample consisting of every second member of the original sample. Such a sample is reasonably large (2,500) and according to our original hypothesis is generated by a sequence of independent identically distributed random variables. Thus had we applied our elementary test again to this subsample, the result would have been a falsification. We can generalize this to any probability hypothesis involving a sequence of random variables which are postulated to be (a) independent and (b) identically distributed. Suppose we test such a hypothesis by collecting a finite initial sequence of observed values of the random variables and calculating a statistic from this sample in the usual way. We can obtain a new test by taking any sufficiently large subset of the original sample and repeating the procedure of the original test. In this way we obtain a number of tests which taken together are much more severe than the original test and in particular test the independence assumptions involved.

Unfortunately this method of obtaining severe tests is, apparently at least, liable to just the same objection which arose when we were discussing random sequences in the context of the frequency theory. Suppose the distribution associated with our hypothesis is the elementary one $P(1) = p$; $P(0) = 1 - p$. Suppose our sample contains n elements. Let us select the subset of this sample which contains just the 1's. Let us further assume that the number n' of elements in this sample is large enough for an application of our methodological rule for neglecting small probabilities to be applicable. It is clear that the resulting test applied to this subset will falsify the original hypothesis. In this kind of way we can arrange for any probability hypothesis to be falsified be it ever so true. A critic might with some justice say: 'Applying your methodological rule for neglecting small probabilities systematically we succeed in falsifying any hypothesis whatever the facts. Your rule certainly and your whole approach almost certainly is thereby discredited.' However, let us not give way to despair but rather pursue the matter a little further.

The construction of such paradoxical tests is evidently made possible by the small chance of falsifying a hypothesis when it is in fact true. Any subsequence could consist of nothing but 1's, but the probability of this is small and the possibility lies in the tail of the distribution.[1] Accordingly we neglect the possibility. For at least one sequence, however, our neglect will lead us astray. The paradoxical test is constructed by picking out this sequence.

Since this problem is the analogue of a problem in the frequency theory we could attempt to deal with it by the methods developed there. These methods, however, appear to be less successful. We can no longer follow Church and appeal to recursive gambling methods. Now we are selecting a subset of a finite sample and any such subset can be specified recursively.

This problem is not a merely academic one but has to be taken into account in practical statistical investigations. This is shown by the following lengthy quotation from Cochran and Cox (1957, pp. 73–4):

> In order that F- and t-tests be valid, the tests to be made in an experiment should be chosen before the results have been inspected. The reason for this is not hard to see. If tests are selected *after* inspection of the data, there is a natural tendency to select comparisons that appear to give large differences. Now large apparent differences may arise either because there are large real effects, or because of a fortuitous combination of the experimental errors. Consequently, in so far as differences are selected just because they seem to be large, it is likely that an undue proportion of the cases selected will be those where the errors have combined to make the differences large. The extreme case most commonly cited is that of the experimenter who always tests by an ordinary t-test, the difference between the highest and lowest treatment means. If the number of treatments is large, this difference will be substantial even when the treatments produce no real differences in effect.

[1] The significance of this last clause will be made clear by the discussion of our falsifying rule in Part III, Chapter 9.

Here Cochran and Cox point out that if we select subsets of the data that exhibit 'large differences' and apply standard tests to these, we will get apparent but not real falsifications. They suggest, in a manner reminiscent of von Mises, that we should design our tests before examining the data. This seems to be an excellent stipulation which solves the present problem. We objected to von Mises' similar solution to the frequency theory problem of randomness on the grounds that it brought in epistemological considerations where mathematical ones were appropriate (see pp. 82–4). However, this objection is no longer applicable. We are not trying to give a mathematical definition of randomness and prove its consistency. Rather we are concerned with the practical problem of designing tests for randomness, and in this practical problem epistemological considerations are certainly relevant. To this general solution some further points can be added. To begin with the use of recursive gambling systems is more help than it might prima facie appear to be. In general, when we test an assumption of independence using a gambling system g, we are in effect comparing the postulate of independence with a postulate of a particular kind of dependence. The opposing postulate is that there are such relations between the trials that 1 will occur more often on those selected by g than on the whole sequence. If we have good reason to suspect that the success of g was fortuitous in a particular finite sequence of observations then we can always continue the sequence of observations and see if g continues to work. Now suppose we have a sample of n repetitions (possible results 0 and 1). We cunningly construct a recursive method g_0 which selects those places at which 1 actually occurs in the sample. Here we evidently have good reason to suspect that g_0's success was fortuitous; and indeed if our original hypothesis was really correct, then g_0 will fail to be successful after the initial segment of n observations.

What is in effect a particular case of this should help to clarify the matter. Suppose the 1's in the original sample of n occurred at say the 1st, 4th, 5th, 7th, ... places; and we construct a bogus test by selecting just those places. This test will of course falsify our original hypothesis but now it is worth noting that the results of a single scientific test are never conclusive. In physics, for example, a single test can reveal a 'stray effect' which never appears on subsequent repetitions of the test. If we suspect that we are dealing with a stray effect we can always repeat the test a number of times. If the phenomenon never reappears, we would disregard it as being merely an oddity. The situation is just the same in the probabilistic case we are dealing with here. We can repeat our test by taking another sample of n and again selecting a subsample consisting of the 1st, 4th, 5th, 7th,... places. If our original hypothesis was correct, this repetition of the test and indeed subsequent repetitions of the same kind will corroborate rather than falsify it. We can thus neglect the result of the bogus test as being a stray effect. These considerations defend our approach of neglecting small probabilities in certain circumstances against the stated difficulty. The discussion also shows how independence assumptions can be tested in practice and thus helps to answer the query which Kolmogorov raises in his monograph when he says (1933, p. 9):

> In consequence, one of the most important problems in the philosophy of the natural sciences is – in addition to the well-known one regarding the essence of the concept of probability itself – to make precise the premises which would make it possible to regard any given real events as independent.

Once again we can quarrel slightly with the formulation. It is not a question of establishing premises from which the independence of events can be deduced, but rather of devising methods of testing an assumption of independence. Such tests of independence are, if I am not mistaken, carried out using gambling systems. The only point worth adding is that there are in addition tests of independence which do not depend – at least in any direct sense – on a gambling system. One such test was used by Kendall and Babington Smith (1939b) on their sets of putative random digits. They describe it thus: 'The lengths of gaps between successive zeros were counted and a frequency distribution compiled for comparison with expectation. This test is called the *gap* test' (p. viii).

(iii) A practical example

Having talked a great deal about testing out probabilistic assumptions, it seems appropriate to give the results of a simple experiment which I carried out. It consisted of tossing an ordinary penny 2,000 times and noting the sequence of heads and tails obtained. The hypothesis was of course that the repeatable conditions of tossing $\mathfrak{S}_s$ together with the set of outcomes $\Omega = (\mathrm{H},\mathrm{T})$ formed a probability system $(\mathfrak{S}_s, \Omega, \mathfrak{F}, p)$ where $p(\mathrm{H}) = p(\mathrm{T}) = \frac{1}{2}$. If we take any subsequence of the 2,000 tosses of length n and this contains m heads, then by equation (1) of Chapter 5, we should have $|m/n - \frac{1}{2}| \leqslant \lambda_\pi/2(n)^{1/2}$. The value of λ_π chosen was 2·21 which corresponds to neglecting a probability of 2·7 per cent. For various reasons which will be given in Part III this seems to me about the correct level. We therefore have the prediction

$$|m/n - \tfrac{1}{2}| \leqslant 1\!\cdot\!105/(n)^{1/2}. \qquad (2)$$

This could be first checked by taking $n = 2,000$, i.e. the whole sequence of tosses. However, as we have pointed out, it is desirable that this test should be supplemented by others more specifically directed at the independence assumptions (the axiom of independent repetitions). These tests were formed by selecting some subsequence of the original sequence of 2,000 tosses using a gambling system, and checking that the relative frequency of heads in this subsequence still satisfies the relation (2). Altogether ten gambling systems were employed. First of all, every second toss was selected $(g(2))$. This system could be started at the first toss $(g(21))$ or at the second $(g(22))$. Next, every fourth toss was selected. This gave in a similar way four gambling systems $g(41), g(42), g(43)$ and $g(44)$. Then the sequences of results were noted which followed a single head $(g(\mathrm{AH}))$ standing for g(After Head), a single tail $(g(\mathrm{AT}))$, two heads $(g(\mathrm{AHH}))$ and finally two tails $(g(\mathrm{ATT}))$. For each gambling system we record in turn (see Table 1 below) the number of members of the corresponding subsequence, the deviation of the relative frequency from 0·5 which by (2) is allowable, the observed relative frequency of heads, and the difference between the observed relative frequency and 0·5. If the observed difference is within the allowable deviation the hypothesis is corroborated. If not, it is falsified. As can be seen all tests resulted in corroboration.

This experiment though very simple is nonetheless instructive. First of all it is an excellent illustration of the thesis that probability theory is a science. From the probability hypothesis we

Table 1

Gambling system	No. of observations	Allowable deviation	Relative frequency of Heads	Difference between r.f. and 0·5
None	2,000	±0·025	0·487	−0·013
g(21)	1,000	±0·035	0·470	−0·030†
g(22)	1,000	±0·035	0·504	+0·004
g(41)	500	±0·049	0·488	−0·012
g(42)	500	±0·049	0·510	+0·010
g(43)	500	±0·049	0·452	−0·048†
g(44)	500	±0·049	0·498	−0·002
g(AH)	974	±0·035	0·505	−0·005
g(AT)	1,025	±0·035	0·470	−0·030†
g(AHH)	487	±0·050	0·503	+0·003
g(ATT)	542	±0·047	0·482	−0·018

infer that the relative frequency in a certain sequence of results should lie in a certain interval. The relative frequency is then calculated from the observations made. As far as I can see, there can be no '*a priori*' or 'logical' reason why the observed frequency should agree with the prediction. If it does agree, the hypothesis is corroborated. If there is a divergence, and we are honest, we should regard the hypothesis as falsified. The empirical nature of probability is here shown in a clear fashion. It is also worth noting that the convergence of the relative frequency to the probability is not *too* fast. In the three cases marked with a † the observed difference is comparable to the allowable deviation. This suggests that the $n^{-1/2}$ law of convergence is indeed correct. Once again we can emphasize how much more precise this is than remarks about 'silent assumptions of fast convergence'.

We also give in Table 2 below the results of some older experiments carried out by Buffon and Karl Pearson. These are quoted from Gnedenko (1950, p. 55).

Once again the results obtained corroborate the theory. This time, however, as we have only the relative frequency in the whole sequence, we cannot test the independence assumptions by applying gambling systems. These tests are therefore less satisfactory than ours, although longer sequences of tosses were used. This example refutes the remarkable view sometimes

Table 2

Author	No. of observations	Allowable deviation	Relative frequency of Heads	Difference between r.f. and 0·5
Buffon	4,040	±0·017	0·507	+0·007
Karl Pearson	12,000	±0·010	0·502	+0·002
Karl Pearson	24,000	±0·007	0·501	+0·001

expressed that for the purposes of probability theory it suffices to record just the relative frequencies obtained in experiments. Of course this is quite wrong. We have to examine the order of the results obtained to see whether they exhibit the requisite randomness or not.

(iv) Definition of random sequences and their generation in practice

So far in this chapter we have operated implicitly with a notion of random sequence. Let me now give an explicit definition.[1] For simplicity I will confine myself to sequences of 0's and 1's. A sequence of 0's and 1's is called a *random sequence* if the ith member of the sequence is a value of the random variable ξ_i where the random variables $\xi_1, \xi_2, \ldots, \xi_n, \ldots$ are independent and have the same distribution given by $p(\xi_i = 1) = p$, $p(\xi_i = 0) = 1 - p$ with $0 \leqslant p \leqslant 1$. This definition includes the cases $00 \ldots 0 \ldots$ and $11 \ldots 1 \ldots$ as degenerate random sequences. This consequence also held in the old von Mises approach and is harmless. It is worth noting that this definition introduces a great mathematical economy over the old frequency approach. In the frequency theory random sequences or collectives were defined by the invariance of limiting frequencies w.r.t. a set of gambling systems. However, when it came to considering the combination of collectives, von Mises had to define and use a notion of *independent* collectives. Thus he introduced two ideas, 'randomness' and 'independence', which were quite differently defined, although it is clear that these two notions are really one and the

---

[1] The following definition has also been suggested by Hacking (1965, pp. 119–20).

same. This fact is shown in the above definition which in effect collapses the notion of randomness into that of independence, and thereby introduces a great economy in the mathematical development.

Another advantage of this definition of randomness is that it ties in very nicely with the way in which random numbers are produced in practice. The method used is this. An experiment $\mathfrak{E}$ is designed in the hope that it will satisfy the hypothesis H that (i) repetitions of $\mathfrak{E}$ are independent, and (ii) the possible results are 0, 1,..., 9 and these have equal probability. A sequence of results of $\mathfrak{E}$ is recorded and these data are used to obtain a number of tests of H. If H passes these tests, the sequence can be accepted as a sequence of random digits. An example of this procedure is provided by Kendall and Babington Smith's tables of random sampling numbers produced in 1939. These authors used a kind of improved roulette wheel. It consisted of a disc divided into ten equal sections marked 0, 1,..., 9, and rotated by an electric motor. From time to time the disc was illuminated so that it appeared stationary, and the number next to a fixed pointer was noted. (For a more detailed description of the randomizing machine see Kendall and Babington Smith, 1939a, pp. 51–3.) A sequence of 100,000 digits was collected using this machine, and this sequence, together with certain of its subsequences, was then subjected to four different kinds of tests. One type of test was the 'gap' test already mentioned. Another was a simple frequency test which consisted in comparing the observed relative frequencies in a subsequence with their expected values, i.e. 1/10. The whole sequence was arranged first in 100 blocks of 1,000 digits, then in 20 blocks of 5,000 digits, and finally in 4 blocks of 25,000 digits. Three of the tests were applied to each of the blocks of 1,000 digits and the four tests to the remaining blocks and to the whole sequence. Out of these 400 tests there were only 6 failures. Four of the blocks of 1,000 digits failed to pass one of the tests and one failed to pass two tests. Even here, however, the divergence from expectation was not very great.

These failures did not cause Kendall and Babington Smith to reject the hypothesis of randomness. Rather they argued that in such a large number of tests it is likely from the nature of statistical testing that there will be a few failures. Nearly all the evidence supports the randomness assumption, and the anomalies can therefore be dismissed as stray effects. This reasoning seems entirely valid and in agreement with our earlier discussion. Kendall and Babington Smith also add that the blocks of 1,000 digits which failed at least one test are probably not very suitable for use in certain practical situations. I will return to this point in a moment. More recently (1955) the Rand Corporation published *A Million Random Digits*. The practical means used to obtain these were different but the underlying principles remained the same. The physical basis of the experiment in this case was the random emission of electrons in certain circuits.

It is interesting to observe that in both the cases mentioned there were considerable practical difficulties in eliminating bias and dependencies. Let us consider Kendall and Babington Smith first. They took readings from the machine themselves and also got an assistant to take some readings. They found, as I have already mentioned, that their own readings satisfied nearly all tests for randomness. However, the assistant's readings showed a significantly higher frequency of even numbers than odd numbers. Kendall and Babington Smith concluded that he must have had a strong unconscious preference for even numbers, and that this caused him to misread the results. The Rand Corporation ran into troubles of a different kind. To make use of the random emission of electrons, it is necessary to amplify the signal. Now the amplifying circuits have a certain 'memory', and this is liable to introduce dependencies even if the underlying emissions are genuinely independent. These examples are highly instructive and also encouraging, for they show that our statistical tests do really enable us to detect biases and dependencies so that we can eliminate them.

It is now an appropriate moment to say a few words about a difficulty which arises when random numbers are used in practice. Suppose we have a sequence of digits which is random in the sense already defined, i.e. produced by the independent repetitions of an experiment whose results are 0, 1, 2, ..., 9 and have equal probability. If the sequence is sufficiently long then there will be a very high probability of having a subsequence of, say, 100 consecutive 0's. Indeed the whole sequence would fail a number of tests for randomness unless such a subsequence appeared. But now suppose we are using the random numbers in practice – say to obtain a random sample of size hundred. Then the subsequence of 100 consecutive 0's would be most unsuitable. In other words a sequence of random numbers may not be suitable for use in practice. Kendall and Babington Smith call a sequence of random numbers which is suitable for practical use a set of random *sampling* numbers. They then put the point thus (1938, p. 153): 'A set of Random Sampling Numbers...must therefore conform to certain requirements other than that of having been chosen at random.' The problem is now: what are these further requirements? Our previous discussion suggests a simple answer which agrees with what Kendall and Babington Smith themselves say.

It has already been emphasized that statistical tests are always provisional and that it is always possible to reject apparent falsifications as 'stray effects' in the light of subsequent evidence. Thus it is perfectly possible to have a sequence which is in fact random but which fails a number of standard tests for randomness. The hundred consecutive 0's just mentioned would be an example of this. Consequently, it makes sense to require that a sequence should not only *be* random, but also satisfy certain standard tests for randomness. Such sequences, I claim, are the ones which are most suitable for practical use.

(v) Relation to randomness as defined in the frequency theory

We now come to the question raised earlier of whether the old frequency approach to randomness retains any significance within the present account. The answer to this must in general terms be 'yes'. The ideas of gambling systems and invariance of frequency play an important rôle in our account of how independence assumptions can be tested. But what of the specific results obtained by the frequentists – say Wald's theorem that given a denumerable set of gambling systems there exists a continuum infinity of random sequences? Do these retain any importance?

I believe that they do and will now explain why. To do so we will make use of an idea of Copeland's. When setting up a form of the frequency theory he says (1928):

It is impossible to devise a physical experiment to test the validity of these assumptions since we are necessarily restricted to a finite number of trials of the event.... Thus our problem is to construct a model universe of events to test the validity of the fundamental assumptions of the theory of probability.

Now of course I would not agree with the first part of this, as I have argued that we can test the assumptions in the finite observable case. Nonetheless it seems to me interesting to construct a model mathematical universe to test out the assumptions in the infinite case. This can be considered as a kind of 'thought-experiment' on the axioms of probability theory. In the model we have corresponding to our finite sequences of observations infinite sequences of 0's and 1's. Corresponding to our practical tests of independence which use gambling systems we have the procedure of selecting infinite subsequences and seeing whether the limiting frequency remains constant. We can now ask: do there exist infinite sequences of 0's and 1's which would satisfy every test we can devise for discovering random sequences, i.e. for discovering sequences of results of independent identically distributed random variables. The results of the frequency theorists show that provided we limit the allowable tests to those which are 'effective' or 'recursively specified', we have a continuum infinity of random sequences with any assigned limiting frequency. This shows that the rather strange and counter-intuitive notion of randomness is in fact quite consistent from a mathematical point of view. It is thus not unreasonable to postulate that randomness occurs in the world. Indeed the result might lead us to the metaphysical speculation that randomness should be much more frequent than order in the world, and that apparent order may well be due to some underlying random process. This speculation has been partly borne out by the development of physics during this century. However, it is in many ways unacceptable to physicists, as is shown by the numerous attempts to explain away statistical portions of physics by an underlying deterministic theory. From the present point of view it would seem more sensible to try to explain the remaining deterministic portions of physics by means of an underlying statistical theory. There have been a few attempts to do this for gravitational theory, but none has as yet been successful.

(vi) Answer to an objection of Braithwaite's

There remain a couple of points connected with randomness and the axiom of independent repetitions to be dealt with. It is worth noticing that according to the frequency theory probability was connected with *ordered* sequences of observations. Now we have replaced collectives $\mathfrak{C}$ by 'repeatable conditions' $\mathfrak{S}_s$, but the requirement of order remains. The spacing conditions must be such that on repeating $\mathfrak{S}_s$ we get an *ordered* sequence: a first, second, third repetition etc. This demand for an ordering has been criticized by Braithwaite. He writes (1953, p. 125):

> ...the notion of limit is an ordinal notion.... But any notion of order seems quite irrelevant to the scientific notion of probability in general, which is concerned with the significance of statements like '51 per cent of children born are boys' in which there is no reference to any order whatsoever.

Actually his example is not a good one to illustrate the point since there is a natural order for births, viz. the time order. Suppose, for example, we confine ourselves to the maternity ward of a definite hospital. Let us observe the sex of the child produced in successive births, and write '1' for a boy and '0' for a girl. In this way we would naturally obtain an ordered sequence just as we do when tossing a coin.

However, other examples can be given where there is no natural order and we have to impose one arbitrarily in order to apply probability theory. These examples occur where the spacing parameter s in $\mathfrak{S}_s$ is literally to do with spatial distances. Suppose, for example, we are considering the molecules of a gas at a particular time t. Our repeatable conditions specify that we must select a particular molecule (the outcome might be its instantaneous velocity at t). Repetitions of the conditions are obtained by taking different, i.e. spatially distinct, molecules. Now evidently there is no natural ordering of the molecules and we must impose one arbitrarily to get our ordered sequence of repetitions. This might seem a dubious procedure but provided the axiom of independent repetitions holds it is easily shown to be legitimate. Since the observations are independent it does not matter what order we take them in. If it is convenient mathematically to impose a certain order, we are quite entitled to do so.

(vii) Probability theory and determinism

Our next point concerns the relations between probability theory and the metaphysical theory of complete determinism.[1] If our general approach is right, and if we can correctly assert that at least one probability system exists in reality, then it follows that at least some objective randomness exists in the world, and so complete determinism is shown to be false. Against this argument, it could be held that we do not in fact ascribe any objective randomness when we apply probability theory to the world, and that probability theory is quite compatible with determinism. I will now examine this line of thought.

The argument against the need for postulating objective randomness is a simple one. Consider tossing a coin. Surely, it could be said, if we knew the exact initial velocities given to the various parts of the coin, the air resistance and the manner in which the coin is allowed to fall, then using Newtonian mechanics we could calculate the result of the toss. The coin is a completely determined Newtonian mechanism. We only use the theory of probability because we don't know the exact initial velocities, and even if we did know them, it would be too tedious to work out the problem exactly. We are content to use the approximate method of probabilities and frequencies. Thus an application of the theory of probability does not involve postulating any objective randomness. We can assume that the process is completely deterministic and use probability theory only because it would be too complicated to work out the result exactly.

---

[1] Metaphysical determinism is a rather vaguely defined theory. For the purposes of the present discussion we will take it as involving the following assumptions. There could be a language $\mathfrak{L}$ for describing the basic structure of the world. This would include concepts of absolute space and time in the Newtonian sense. We could in principle give a total world description D_t at any time t. This might consist, say, of an account of all the particles, their positions, velocities and interactions. Further, we would have laws of nature which would enable us, given a finite subset of D for some fixed $t = \tau$, to calculate (i) the rest of D_τ and (ii) the whole of D_t for any other $t > \tau$. These calculations could be carried out recursively. The world described by $\mathfrak{L}$ is then said to be deterministic for $t > \tau$. This formulation seems very complicated, but anyone who tries to give an adequate and reasonably precise characterization of metaphysical determinism will appreciate the difficulties involved.

This argument is plausible but omits one important point. It may be true that given the exact initial conditions of a particular toss its result is thereby determined, but the exact initial conditions of tossing will *vary randomly* from one toss to the next. Thus the argument does not eliminate objective randomness but (perhaps) shows that the random variation of heads and tails can be explained in terms of the random variation of the initial conditions.

(viii) Discussion of some arguments of Khintchine's

This is a convenient point to discuss an interesting paper of Khintchine's ('The method of arbitrary functions and the struggle against idealism in probability theory', 1952), for Khintchine uses a variant of the above argument in order to criticize von Mises. Khintchine begins by pointing out that von Mises' frequency theory is based on Mach's general theory of knowledge. This in turn is an idealistic position and was criticized, decisively he claims, by Lenin from the materialist point of view. These points are of course in entire agreement with my own approach (cf. Introduction, p. 32) except that I have criticized Mach in terms of Popper's realistic philosophy of science rather than in terms of dialectical materialism. *In this particular case*, however, the two criticisms come to very much the same thing.

I am in agreement then with Khintchine's basic approach but he proceeds to develop the argument in a different (and in my view incorrect) manner. He says (1952, p. 267):

> If we adopt the materialist [i.e. in Popperian terms 'realist'] point of view, we proceed in the firm certainty that underlying the experimental results which we continually observe, there are in objective nature quite definite causes which govern the development of the process in conformity to law, and that the results can and will be explained in the course of further scientific progress, if we start from the properties of these causes.

I have argued from the realist point of view that probabilities should be taken as standing in their own right for things which exist in reality rather than as instances of a concept to be operationally defined in terms of observables. However, Khintchine takes the realist position as one requiring that we should be able to explain the sequences of experimental results in terms of 'definite causes' existing 'in objective nature'. Now the word 'cause' is here rather ambiguous. If a fixed probability can be considered as a 'definite cause' then I am in complete agreement with Khintchine; but if he means by his use of this phrase that we should try to explain probability theory by an underlying deterministic theory then I disagree. Such a programme is not demanded by realism because there is no reason why we should not regard probabilities (like 'forces' and 'masses') as real. Further, an underlying deterministic mechanism would refute a probabilistic theory by contradicting the randomness which, as I have argued, is demanded by such a theory. Khintchine tries to give an explanation (at least in outline) of certain probabilistic phenomena in terms of underlying causes. If we examine this account, there appear to be gaps in it which, if filled, lead us back to the very frequency theory which is being criticized. Nonetheless Khintchine's results do seem valuable if given a rather different interpretation from the one he accords them.

Khintchine considers in detail the case of a roulette wheel.[1] He supposes that the ball is given an initial velocity v and is stopped after a time t. The result is either red denoted by 'A' or black denoted by 'B'. t is considered as fixed from one trial to the next so that the result depends only on v. v is therefore supposed to vary from trial to trial. Khintchine next assumes as an idealization that we are dealing with an infinite sequence of trials and the limiting frequency of those in which v lies between v_1 and v_2 is given by $\int_{v_1}^{v_2} f(v)\,dv$ where f is an arbitrary function s.t. $f(v) \geqslant 0$ $(0 \leqslant v < +\infty)$; $\int_0^\infty f(v)\,dv = 1$. He then calculates that

$$p(A) = \sum_{k=0}^{\infty} \int_{k/nt}^{(k+1/2)/nt} f(v)\,dv$$

$$p(B) = \sum_{k=0}^{\infty} \int_{(k+1/2)/nt}^{(k+1)/nt} f(v)\,dv$$

---

[1] His treatment of this case is very similar to that given by Poincaré (1902, pp. 201–3).

10

and finally shows that for arbitrary f, $p(A)$ and $p(B) \to \frac{1}{2}$ as $t \to \infty$.

The trouble with this approach is that it says nothing about randomness. For us to have a genuine probability of A, it is not enough that A occurs with a certain frequency in some sequence of repetitions. The occurrence of A must also be random within that sequence. In order to get this randomness we must postulate that the variations of v are also random; but then our hypotheses about v are really equivalent to saying that on a trial of the roulette wheel

$$\text{Prob}\,(v_1 \leqslant v \leqslant v_2) = \int_{v_1}^{v_2} f(v)\,dv$$

where the probability here is analysed in terms of von Mises' frequency theory. From this we can calculate the probability of A, but this probability must, in turn, be analysed in terms of the frequency theory.

It seems then that if we fill out Khintchine's argument we are led to an analysis of probability in terms of von Mises' frequency theory. Khintchine would, I think, deny this. He ascribes to the upholders of the frequency theory the following very strong position (1952, p. 267):

> The stable frequency of 1/6 for the throwing of a 5 has been determined by many experiments; in probability theory this result is taken as the foundation for further calculations; all consideration of the causes of this result lie outside the 'positive' science.

Now Khintchine's arguments do refute this form of the frequency theory but not a more liberal form. We could, for example, retain the limiting frequency definition of probability and the frequency theory approach to randomness and yet following Khintchine explain macroscopic random variations of 'red' and 'black' in terms of microscopic random variations in initial conditions. Further, the Khintchine explanation of macroscopic by microscopic will still hold *whatever* account of probability we give. I conclude that Khintchine's arguments do not contribute (as he claims) to a philosophical analysis of probability, but they do contribute to a scientific understanding of random phenomena.

This last contribution seems to me most valuable. If Khintchine's deduction is correct, it shows that we can regard macro-random processes as an amplification of micro-random processes. More generally we can speculate that all macro-randomness is an amplification of randomness at the atomic level. This is certainly true of the method of obtaining random numbers which consists of amplifying the random emission of electrons in certain circuits. It could be that this is a simplified case of the general way in which randomness is produced. Consider, for example, random variations of say 'height' in certain biological populations. Height is controlled by certain genetic mechanisms and thus we are again driven back to atomic considerations. However, to reduce all randomness to atomic randomness is not to dispose of it altogether.

(x) Possible modifications of the axioms of probability in the light of experience

I will close this chapter with a discussion of the following question[1]: Might we ever modify the basic axioms of probability in the light of recalcitrant experience? If probability theory is indeed a science it should in principle be possible to alter its axioms because of certain experimental findings – but could this ever in fact occur? This question must be carefully distinguished from the problem of whether individual probability hypotheses can be falsified. Consider for example the coin experiment described earlier. Here the individual probability hypothesis was that the repeatable conditions of the tossing with outcomes (H, T) and distribution $p(H) = p(T) = \frac{1}{2}$ formed a probability system. This hypothesis could, as we saw, only be falsified by frequency evidence if it was agreed, as a methodological rule, to neglect small probabilities in certain circumstances when testing. Let us suppose from now on that such a rule has been adopted. The individual hypothesis is then certainly falsifiable. As we saw we can predict that the relative frequency of heads should lie within certain limits and this prediction can be compared with the observed frequency. But does this mean that the axioms of probability might be altered in the light of experience? Not necessarily. Suppose for example our coin

---

[1] I have greatly profited from discussions of this question with my friend M. O. Hill – though our views remained in the end different.

hypothesis had been falsified. Would we regard this as over-
throwing the axioms of probability? Of course not. We would
almost certainly have tried some different probability hypothesis.
For example, we might have postulated that prob(heads) =
$p > \frac{1}{2}$, or that the sequence of tosses was really a Markov chain
with certain weak 'after-effects' operating. It could therefore
be argued that, although individual probability hypotheses can
be falsified, the axioms of probability will never be changed in
the face of recalcitrant experiences. I do not agree with this
conclusion and will now argue against it.

My main point is that Poincaré in 1902 advanced arguments
of just the above kind to show that Euclidean geometry and
Newtonian mechanics would never be altered in the light of
experience. Yet only fifteen years after he wrote, both Euclidean
geometry and Newtonian mechanics had been changed in order
to explain certain experimental findings. I see no reason why
the axioms of probability should not suffer a similar fate.

It will be instructive to quote the famous passage from
Poincaré where he puts forward this argument. He says (1902,
p. 72):

> If Lobatchewsky's geometry is true, the parallax of a very
> distant star will be finite. If Riemann's is true, it will be
> negative. These are results which seem within the reach of
> experiment, and it is hoped that astronomical observations
> may enable us to decide between the two geometries. But what
> we call a straight line in astronomy is simply the path of a ray
> of light. If, therefore, we were to discover negative parallaxes,
> or to prove that all parallaxes are higher than a certain limit,
> we should have a choice between two conclusions: we could
> give up Euclidean geometry, or modify the laws of optics, and
> suppose that light is not rigorously propagated in a straight
> line. It is needless to add that every one would look upon this
> solution as the more advantageous. Euclidean geometry,
> therefore, has nothing to fear from fresh experiments.

This is correct except for the last two sentences. In fact it turned
out to be more advantageous to give up Euclidean geometry.

Now consider probability theory. Let us suppose (a not
improbable assumption) that some peculiar result is recorded in
the domain of particle physics. This can be satisfactorily

explained by altering the probability axioms which appear implicitly in quantum mechanics. Suppose further it is accepted that all macro-randomness is an amplification of micro-randomness. (I have argued for this thesis above.) Then we would correspondingly have to alter the probability axioms at the macro-level as well.

Two points should perhaps be made about this. First of all, even if a new probability theory were introduced in the way we have described, it would certainly, in most practical cases, give results which agreed to a high degree of approximation with those obtained from classical probability theory. The general principle of correspondence discussed in Part I ensures that this would be so. Thus we would no doubt continue to use classical probability theory in most cases, just as we now often use Newtonian mechanics and Euclidean geometry as approximations.

Secondly, the physicists were only able to give up Euclidean geometry because pure mathematicians had developed non-Euclidean geometries. This suggests that it might be useful for pure mathematicians to try to develop 'non-Euclidean' probability theories by altering the standard axioms in various ways.

CHAPTER SEVEN

Probabilities of Single Events: Popper's Propensity Theory

(i) Singular probabilities

In this chapter we turn to a problem in the philosophy of probability which, though intriguing, is perhaps of secondary importance. It is the question of whether we can introduce probabilities for single, unrepeated events. The problem is perhaps most easily seen in terms of the frequency theory. We therefore introduce probabilities for attributes which appear in a long sequence of events: an empirical collective. Now does it make sense to ascribe a probability to the appearance of the attribute on a single event of the sequence as well as in the whole sequence? We can, for example, speak validly of 'the probability of heads in this sequence of tosses'. Can we also speak of 'the probability of heads on the next toss' granted that 'the next toss' is an individual event which will never be repeated in the course of nature? Of course we have replaced collectives by sets of repeatable conditions, but, as usual, a corresponding problem arises in the new approach. Granted that we introduce probabilities for a set of repeatable conditions, can we also introduce them for particular instances of these conditions? It is interesting to note that von Mises explicitly denies the validity of such probabilities of single events. The example he considers is the probability of death. We can certainly introduce the probability of death before 80 in a sequence of, say, 40-year-old English non-smokers. But can we consider the probability of death before 80 for an individual person? Von Mises says 'No' (1928, p. 11):

We can say nothing about the probability of death of an individual even if we know his condition of life and health in detail. The phrase 'probability of death', when it refers to a single person has no meaning at all for us. This is one of the most important consequences of our definition of probability....

We may well ask: Is there any point in trying to introduce such singular probabilities (as Popper calls them)? Do they serve any useful function? It appears, perhaps surprisingly, that they do. An example drawn from the field of operational research will illustrate this. Suppose an oil company is considering whether to build a refinery. It may be interested in the probability (in some sense) of the enterprise proving successful. Yet here we may be dealing with a single event. The company may possess only enough capital for one refinery so that there is no chance of the venture being repeated. Repetitions are ruled out on contingent economic grounds. It may further be difficult in this case to produce a satisfactory repetition even granted enormous supplies of capital. Of course it would be easy under this last-stated condition to build another similar refinery, but the existence of the first refinery might well affect the profitability of the second. In other words the repetition would not be independent (the axiom of independent repetitions would not be satisfied). To satisfy the independence condition it might be necessary to build the refinery in an economically isolated part of the world – say in the communist rather than the capitalist area of influence. Such a repetition might be ruled out on political as well as economic grounds. The point is perhaps emphasized by considering the calculations which are doubtless carried out in the Pentagon concerning the probability of total war.

In these cases we have single and unrepeatable events. Yet probability calculations are often and usefully applied. It is up to us to explain how this can be justified from our point of view. This problem is, apparently at least, more easily dealt with on a subjective theory of probability which identifies probability with degree of belief. Evidently different people can believe to different degrees in the proposition that the refinery will be profitable or that there will be total war. But does it make sense to speak of an objective probability of these events? As we are defending the scientific view of probability we must show that it does.

Von Mises, as we have seen, denied the possibility of probabilities of single events. Popper, however, in the original edition of *The Logic of Scientific Discovery* (1934), pointed out that such probabilities could be introduced in the frequency theory in a simple, even trivial, fashion.

Popper there adopts von Mises' view that probabilities are correctly defined within collectives, which he calls 'reference sequences' and denotes usually by α. $\alpha F(\beta)$ stands for 'the probability of the attribute or property β within the reference sequence α'. He goes on to consider probabilities for single events or 'formally singular probability statements', and remarks that (1934, p. 210): 'From the standpoint of the frequency theory such statements are as a rule regarded as not quite correct in their formulation, since probabilities cannot be ascribed to single occurrences, but only to infinite sequences of occurrences or events.' However, he continues:

> It is easy…to interpret these statements as correct, by appropriately defining formally singular probabilities with the help of the concept of objective probability or relative frequency. I use '$\alpha p_k(\beta)$' to denote the formally singular probability that a certain occurrence k has the property β, in its capacity as an element of a sequence α – in symbols: $k \in \alpha$ – and I then define the formally singular probability as follows:
>
> $$\alpha p_k(\beta) = \alpha F(\beta) \qquad (k \in \alpha) \text{ (Definition)}.$$

In other words the probability of a particular outcome occurring at a certain fixed member of a collective equals the probability of that outcome in the collective as a whole.

Now admittedly this definition only ascribes probabilities to single events qua members of a particular collective and a particular event can of course belong to a number of collectives. Thus probabilities are not assigned to single events in an absolute fashion but only conditional to the way the events are described. We will discuss objections based on this point later. For the moment it must be admitted that the definition does introduce singular probabilities albeit in a formal and conditional way. Apparently at least the definition is simple and unexceptionable. It is surprising therefore that a defect exists in this account, a defect which was, moreover, discovered by Popper himself. In his 1959 paper Popper describes his objection and it leads him to replace the frequency theory of probability with what he calls the propensity theory. I will now discuss these matters.

(ii) Popper's propensity theory

Popper's argument is this. Begin by considering two dice: one regular and the other biased so that the probability of getting a particular face (say the 5) is 1/4. Now consider a sequence consisting almost entirely of throws of the biased die but with one or two throws of the regular interspersed. Let us take one of these interspersed throws and ask, what is the probability of getting a 5 on that throw? According to Popper's previous definition this probability must be 1/4 because the throw is part of a collective for which prob(5) = 1/4. But this is an intuitive paradox, since it is surely much more reasonable to say that prob(5) = 1/6 for any throw of the regular die.

One way out of this difficulty is to modify the concept of collective so that the sequence of throws of the biased die with some throws of the regular die interspersed is not a genuine collective. The problem then disappears. This is just what Popper does (1959, p. 34):

> All this means that the frequency theorist is forced to introduce a modification of his theory – apparently a very slight one. He will now say that an admissible sequence of events (a reference sequence, a 'collective') must always be a sequence of repeated experiments. Or more generally he will say that admissible sequences must be either virtual or actual sequences which are *characterized by a set of generating conditions* – by a set of conditions whose repeated realization produces the elements of the sequences.

He then continues a few lines later:

> Yet, if we look more closely at this apparently slight modification, then we find that it amounts to a transition from the frequency interpretation to the propensity interpretation.

The 'table of differences' between von Mises and Kolmogorov given in Chapter 4 proves helpful here. The first item stated that whereas von Mises was prepared to define probabilities on empirical collectives, Kolmogorov wanted to introduce them on certain repeatable conditions ☾. Kolmogorov may have been influenced in this by Tornier's paper, 'Grundlagen der Wahrscheinlichkeitsrechnung', which he mentions in his bibliography. Tornier there introduces what he calls an 'experimental rule' (*Versuchsvorschrift*). Probabilities are then associated with the 'logically possible realizations' of such experimental rules, and of course with sets of such possible realizations. However neither Kolmogorov nor Tornier gives any arguments for the change from von Mises' collective concept. What Popper has done is to supply an argument: an argument which I regard as correct and highly valuable. My only slight disagreement with Popper is that he sometimes formulates his point in Tornier-like fashion in terms of *experimental* conditions. Thus he defines the propensity interpretation of the equation '$p(a, b) = r$' as follows (1957a): 'r measures the tendency or propensity of *the experimental set-up b* to produce, upon repetition, the result a' (my italics). However, the repeatable conditions may not be experimental conditions if 'experimental' is interpreted in a strict sense. For example, we often consider the probability that a molecule of a certain gas has a velocity between v and $v + u$; but we do not literally observe a molecule of the gas experimentally and note its velocity. I therefore prefer Kolmogorov's looser formulation.

There is, however, rather more to Popper's notion of propensity than is involved in the change from collectives to conditions. The word 'propensity' suggests some kind of dispositional account, and this marks a difference from the frequency view. A useful way of looking into this matter will be to consider some earlier views of Peirce which were along the same lines. These are contained in a passage first published in 1910. The quotation comes from the reprint, *Essays in the Philosophy of Science* in the American Heritage Series (p. 79). Part of it is quoted in Braithwaite (1953).

I am, then, to define the meaning of the statement that the *probability*, that if a die be thrown from a dice box it will turn up a number divisible by three, is one-third. The statement means that the die has a certain 'would-be'; and to say that the die has a certain 'would-be' is to say that it has a property, quite analogous to any *habit* that a man might have. Only the 'would-be' of the die is presumable as much simpler and more definite as the die's homogeneous composition and cubical shape is simpler than the nature of the man's nervous system and soul; and just as it would be necessary, in order to define a man's habit to describe how it would lead him to behave and upon what sort of occasion – albeit this statement would by no means imply that the habit *consists* in that action – so to define the die's 'would-be' it is necessary to say how it would lead the die to behave on an occasion that would bring out the full consequence of the 'would-be'; and this statement will not of itself imply that the 'would-be' of the die *consists* in such behaviour.

Peirce then goes on to describe 'an occasion that would bring out the full consequence of the "would-be"'. Such an occasion is an infinite sequence of throws of the die and the relevant behaviour of the die is that the appropriate relative frequencies fluctuate round the value 1/3, gradually coming closer and closer to this value and eventually converging on it. Nothing is mentioned about 'excluded gambling systems'.

Peirce is of course mistaken in speaking of the 'would-be' as a property of the die. Obviously it depends on the conditions under which the die is thrown, as is shown by the following two interesting examples of Popper's. Suppose first we had a coin biased in favour of 'heads'. If we tossed it in a lower gravitational field (say on the Moon), the bias would perhaps have less effect and prob(heads) would assume a lower value. This shows an analogy between probability and weight. We normally consider weight loosely as a property of a body whereas in reality it is a relational property of a body w.r.t. in a certain gravitational field. Thus the weight of a body is different on the Moon whereas its mass (a genuine property of the body) is the same. For the second example we can use an ordinary coin but this time, instead of letting it fall on a flat surface, say on a table top, we allow it to fall on a surface in which a large number of slots have been cut. We now no longer have two outcomes 'heads' and 'tails' but *three*, viz. 'heads', 'tails' and 'edge'; the third outcome being that the coin sticks in one of the slots. Further, because 'edge' will have a finite probability the probability of 'heads' will be reduced. This example shows that not only do the probabilities of outcomes change with the manner of tossing but even that the exact nature of the outcomes can similarly vary.

Despite this error Peirce has made what seems to me a valuable point in distinguishing between the probability of the die as a dispositional quantity, a 'would-be', on the one hand and an occasion that would bring out the full consequence of the 'would-be' on the other. The importance of making this distinction is that it allows us to introduce probabilities as 'would-be's' even on occasions where the full consequences of the 'would-be' are not manifested, where in effect we do not have a long sequence of repetitions. On the other hand, if we regard probabilities as 'consisting in such behaviour' then it will only make sense to introduce probabilities on 'occasions of full manifestation', i.e. only for long sequences of repetitions. All this will become clearer if we now return to von Mises and Popper.

It is a consequence of von Mises' position that probabilities ought only to be introduced in physical situations where we have an empirical collective, i.e. a long sequence of events whose outcomes obey the two familiar laws. If we follow the Popper line, however, it is perfectly legitimate to introduce probabilities on a set of conditions *even though these conditions are not repeated a large number of times*. 'The probability of outcome A of conditions $\mathfrak{S} = 1/6$' = 'The conditions $\mathfrak{S}$ have a "propensity" or "disposition" or "would-be" such that, were they to be repeated infinitely often, a random sequence of attributes would result in which attribute A would have the limiting frequency $1/6$'.[1] As the random sequence here is hypothetical, we do not have to consider it as corresponding to a long sequence given in experience. We are allowed to postulate probabilities (and might even obtain testable consequences of such a postulation)

---

[1] This formulation can be compared with what is perhaps Popper's clearest statement of his 'propensity' view (Popper, 1959, p. 35): 'But this means that we have to visualize the conditions as endowed with a tendency or disposition, or propensity, to produce sequences whose frequencies are equal to the probabilities; which is precisely what the propensity interpretation asserts.'

when the relevant conditions are only repeated once or twice. Thus in a deeper sense than the one considered hitherto, the propensity theory opens the way towards the introduction of probabilities for single events. This then is Popper's new theory. I will now attempt to assess it relative to my own position.

(iii) Similarities between Popper's propensity theory and von Mises' frequency theory

So far I have stressed the differences between Popper's propensity theory and a von Mises type of frequency theory. Now we must turn to the similarities. The first point to note is that it would be possible to change from 'collectives' to 'conditions' while retaining the entire mathematical formalism of the frequency theory. One could deal with sequences of attributes $(\omega_1, \omega_2, \ldots, \omega_n, \ldots)$, postulate the axioms of convergence and randomness for such sequences and define the probability of an attribute as its limiting frequency. The only difference from the old view would lie in the interpretation of the sequences. They would no longer be seen as corresponding to long sequences of attributes given in practice (empirical collectives) but as being generated by the actual or hypothetical repetitions of a set of conditions. This last change is an important one, but it does not in my opinion justify saying that we have introduced a new theory of probability when we retain all the mathematical formulations of the old theory. This is seen more clearly if we consider the classification of scientific theories of probability given earlier. I said that a scientific theory of probability is a *frequency* theory if it identifies probability and frequency either in the mathematical formalism or in an informal supplement designed to tie the theory in with experience. A *non-frequency* theory is one in which probability is not identified with frequency and in which the link between the two is established by non-definitional means. It is clear that the change from 'collectives' to 'conditions' is compatible with either a frequency or a non-frequency theory in the above sense.

It is difficult to judge Popper's own position precisely because he has not as yet published a full account of his propensity theory. As far as one can tell from a number of footnotes and side remarks, however, he would adopt a position similar to that sketched in the previous paragraph,[1] i.e. he would retain most of the mathematical ideas of the old frequency theory. This remark needs a little qualification because Popper has never accepted the standard mathematical formulation of the frequency theory but has developed variant accounts of his own. In the original (1934) edition of *The Logic of Scientific Discovery* he only allowed place selections of the form 'select those elements which succeed a finite group of such and such a character', and made use of points of accumulation rather than limits (cf. 1934, Ch. VIII). In 1959, however, he decided to employ the notion of an 'ideally random sequence' (cf. Appendices iv and *vi and certain footnotes to the *Logic of Scientific Discovery*, Ch. VIII). These random sequences are defined using place selections, but in such a way that their definition does not presuppose the existence of limiting frequencies for the attributes. It is the actual frequencies in certain finite initial segments which are invariant under the place selections. We thus do not assume an 'axiom of convergence' for ideally random sequences, but it can be shown that from this strong concept of randomness the convergence of relative frequencies follows. 'Ideally random sequences' have, however, only been defined in the case of equidistribution so that the theory stands in need of a good deal of mathematical development. I may add that Popper has a

[1] (Footnote added in proofs.) I now believe that this is a misinterpretation of Popper's views. My mistake lay in supposing that Popper wished to incorporate his theory of 'ideally random sequences' into his new propensity theory. In fact, however, at the end of Appendix *vi to the *Logic of Scientific Discovery*, where Popper introduces ideally random sequences, he says (Popper, 1959, p. 362): 'Today, some years after having solved my old problems in a way which would have satisfied me in 1934, I no longer quite believe in the importance of the undoubted fact that a frequency theory can be constructed which is free of all the old difficulties.'

We see from this that ideally random sequences are introduced by Popper to enable him to construct a frequency theory 'free of all the old difficulties' but he 'no longer quite believe(s)' in the importance of this undertaking – presumably because he has abandoned the frequency theory with its concept of randomness in favour of a propensity view which does not employ this concept.

I still believe that my comments and criticisms apply validly to the position under discussion. This position is a possible one. However it is not held by Popper, and consequently a discussion of it must inevitably be of less interest.

number of other motives for introducing these 'ideally random sequences', but I shall not discuss these in detail here. First of all, such sequences can be constructively generated, and secondly, he hopes to use them to solve the problem (mentioned earlier) of falsifying probability statements.

These hints lead to the following reconstruction of how Popper might develop his theory in detail. He would begin with the notion of a repeatable set of conditions $\mathfrak{S}$. It would then be postulated that in the circumstances appropriate to the theory of probability the repetition of $\mathfrak{S}$ would produce an 'ideally random' sequence of attributes. Next it would be shown that in such ideally random sequences the limiting frequency of any attribute existed. Finally the probability of an attribute would be defined as its limiting frequency. Thus by postulating a stronger axiom of randomness Popper avoids the axiom of convergence. However, it is clear that this account incorporates many mathematical features of the old frequency theory which are incompatible with the measure theoretic approach, and so unacceptable. First of all, probability is identified with limiting frequency. Secondly, a notion of randomness distinct from independence is introduced. However, it has been shown that these two notions are really one and the same and can, with greater economy, be treated as such. For all these reasons I would prefer to classify Popper's theory as a variant of the frequency theory rather than as a new 'propensity' view.

(iv) A criticism of the propensity theory

But how about our own approach? Should we not perhaps call this a propensity account? After all, we 'deduce' that where we correctly assign probabilities a series of repetitions of the under-lying conditions will produce a random sequence with relative frequencies which converge as $n^{-1/2}$ to the probability. Could we not consider probabilities as propensities to produce such sequences? There is really no harm in such a suggestion, but I am reluctant to adopt it for the following reason. It suggests too strong a connection between probabilities and long sequences of repetitions and obscures the fact that we can often use the probability calculus to obtain results when the underlying conditions are only repeated once. An analogy should help to make this clear. From the law of gravity and certain other assumptions we can deduce that planets move in approximate ellipses, and we could consequently say that gravity is a propensity towards elliptical motion. However, in other conditions we can deduce from the law of gravity that non-elliptical motion will occur. Thus such a 'propensity' view of gravity is a little misleading. In probability theory, if we introduce a probability p of A relative to conditions $\mathfrak{S}_s$, we can deduce that if $\mathfrak{S}_s$ are repeated a large number of times the relative frequency of A will be approximately equal to p. But suppose now the $\mathfrak{S}_s$ are so defined that a single repetition of them is a whole realization of a Markov chain. We might be able to deduce (neglecting small probabilities as usual) that for such a *single* realization the relative frequency of A would differ from p by nearly a half, i.e. nearly as much as possible. A variant of the game of red or blue might produce such a result. In this case we deduce that a certain result will hold on a single repetition of the conditions and, further, the result is that the relative frequency will be entirely different from the probability. The connection between the laws of probability and the appearance of random sequences of repetitions with convergent relative frequencies is of course very important. However, I don't think that the connection is sufficiently close for us to consider probability as *particularly* a propensity towards the production of random sequences.

(v) Another argument for the axiom of independent repetitions
This concludes my discussion of the propensity theory. In the course of introducing this theory Popper used an argument involving singular probabilities; I next want to make use of a similar argument to support the axiom of independent repetitions. The general form of Popper's argument was this. He introduced a plausible definition of probabilities of single events. He then showed that this definition combined with the ordinary notion of collectives led to an intuitive contradiction in some cases. He therefore concluded that we should replace 'collectives' by 'conditions'. We can state Popper's definition of singular probabilities using the idea of repeatable conditions as follows. Let A be an outcome which occurs on the single event E (say a particular toss of a coin where A is heads). Let $p(A, E)$ stand for the singular probability of A on E. Now we claim that such singular probabilities cannot be introduced in an absolute sense

but only relative to a set of repeatable conditions $\mathfrak{S}_s$ where E satisfies $\mathfrak{S}_s$. Let us consequently denote the singular probability by $p_{\mathfrak{S}_s}(A,E)$. We then define it by

$$p_{\mathfrak{S}_s}(A,E) = p(A,\mathfrak{S}_s) \qquad \text{def.} \qquad (1)$$

This corresponds exactly to Popper's frequency theory definition quoted above (see p. 142). I now want to argue that if we combine (1) with the introduction of probabilities on repeatable but dependent conditions we get an intuitive contradiction. This is an argument for limiting the introduction of probabilities to repeatable *and* independent conditions, i.e. for the axiom of independent repetitions. This argument is I think less strong than the one given in Chapter 5, because our intuitions about singular probabilities are less strong than those concerning the connection between probability and frequency. Nonetheless it still has some force.

For a little variety let us consider a new set of repeatable but dependent conditions. We define these as follows:

$\mathfrak{S} =$ take a subset of an ordinary pack of 52 playing cards. Shuffle well and turn over the top card. The name of this card is the outcome of $\mathfrak{S}$.

$s =$ start with the full set of 52 cards. After each repetition remove the top card and continue with the remaining set of cards. When all 52 cards are exhausted start with the whole pack again.

It is clear that $\mathfrak{S}_s$ is a repeatable but dependent set of conditions. Now consider the singular probability of getting say the ace of spades on a particular repetition of these conditions. What value can we intuitively assign to the probability of this single event? Only two possible values are in my opinion admissible: namely 0 if the ace of spades has already appeared and $1/n$ if there are n cards in the pack on that particular go and the ace of spades has not yet appeared. But now suppose, contrary to the axiom of independent repetitions, we introduce prob(ace of spades, $\mathfrak{S}_s$). The only reasonable value for this would be $1/52$. But then according to formula (1) we would have to call the singular probability of getting the ace of spades $1/52$ on each repetition, which contradicts our previous considerations. Once again we have to conclude that probabilities relative to repeatable but dependent conditions are inadmissible.

11

Against this argument two objections[1] could be raised. First it could be maintained that the 'contradiction' arises by confusing two singular probabilities. Suppose, for example, we have turned up to date the king of hearts (C_1) and the two of diamonds (C_2). Let us consider the repeatable conditions which state that we shuffle a pack minus C_1 and C_2 and turn over the top card. Let us denote these repeatable conditions by $\mathfrak{S}_s \wedge C_1 \wedge C_2$. Then we have two singular probabilities: (i) The singular probability of getting the ace of spades if the event is considered qua instance of $\mathfrak{S}_s \wedge C_1 \wedge C_2$. This is 1/50. (ii) The singular probability of getting the ace of spades if the event is considered qua instance of $\mathfrak{S}_s$. This is 1/52. The alleged 'contradiction' comes from confusing these two singular probabilities. To this I reply that only the first of these singular probabilities seems to me to be intuitively admissible. Now it is interesting to note that *this* singular probability (1/50), i.e. the one which most people would at once adopt in these circumstances, is relative to a set of conditions $\mathfrak{S}_s \wedge C_1 \wedge C_2$ which are both repeatable and independent.

The second objection is really a variant of the first. It runs thus. Of course if we know that C_1 and C_2 have come up we naturally assign the probability 1/50. But supposing we don't know this, then the value 1/52 would be reasonable. Further, if we are interested in the probability of the single event, we should not necessarily take into account previous results. My answer to this is that if we don't know the previous results then we ought to say that we just don't know the value of the singular probability for getting the ace of spades. Knowledge of previous results is quite in order in this case. We are only considering the probability of a single event qua member of a sequence of events (repetitions). Thus it is reasonable that we should be allowed to know the results of the previous members of the sequence. But this consideration leads at once to the axiom of independent repetitions. If we are going to assign the *same* singular probability to each member of the sequence, then the results of previous events should be irrelevant. But if previous results are irrelevant then the sequence is one of independent events.

It is interesting to note that von Mises uses a form of the above argument to support his axiom of randomness. His principal reason for introducing this axiom was to enable him to derive the binomial law. However, in *Probability, Statistics and Truth* he offers some further considerations in favour of the axiom. He asks us to imagine a sequence of stones by the roadside. Every tenth stone is large and the intermediate ones are small. Observing these stones in turn we obtain a sequence of the attributes 'large' and 'small' in which 'large' has the limiting frequency 1/10 and 'small' 9/10. However, von Mises does not think it is correct to say that prob (large) = 1/10, because (1928, p. 23): 'After having just passed a large stone, we are in no doubt about the size of the next one; there is no chance of its being large.' This of course is just the argument about singular probabilities. We are not satisfied with prob (large) = 1/10, because prob (large) = 1/10 does not intuitively hold for individual observations. We would rather want to say that prob (large) = 1 or 0 according to our position along the line of stones. Unless we have randomness, the introduction of singular probabilities leads to intuitive contradictions.

(vi) Discussion of a possible simplification
The existence of probabilities of single events suggests a simplification of the theory of probability, but on closer inspection this simplification turns out not to be viable after all. I will now describe it briefly. I have insisted that probabilities should only be defined on repeatable conditions and yet have admitted that it is sometimes possible to test ascriptions of probabilities where the conditions are in fact only instantiated once. Why, therefore, it might be asked, bother with this condition of repeatability? Why not introduce probabilities straight away for particular situations or particular events? After all, even cases where we appear to have a long sequence of tosses of a coin could be dealt with by considering the long sequence as a single event. This suggestion is an attractive one as it would appear to avoid the complicated questions connected with repeatability and thus to simplify the theory. There are, however, fatal objections to it.

The trouble with trying *ab initio* to introduce probability for particular events or particular situations is that intuitively the probability will vary according to the set of conditions which the event is considered as instantiating – according in effect to how we describe the event. But then we are forced to consider probabilities as attached primarily to the conditions which describe the event and only secondarily to the event itself which instantiates these conditions. But the conditions – like any description – will be repeatable. Thus we are driven back to questions of repeatability and independence, and there seems to be no way of avoiding them. I will now argue in detail for the main premiss of this argument, namely that the probabilities of single events vary with the way in which the event is described, and will do so by means of a series of examples. I will then show that this 'conditionality' of singular probabilities does not detract seriously from their utility. Some authors, e.g. Ayer, have thought that this conditionality constitutes a grave difficulty for probability theory, but I will argue that this is not the case.

Any standard example of a singular probability will serve my purposes. Consider for example the probability of a certain man aged 40 living to be 41. Intuitively the probability will vary depending on whether we regard the individual merely as a man or more particularly as an Englishman; for the life expectancy of Englishmen is higher than that of mankind as a whole. Similarly the probability will alter depending on whether we regard the individual as an Englishman aged 40 or as an Englishman aged 40 who smokes two packets of cigarettes a day, and so on. This shows that probabilities must be considered in the first instance as dependent on the properties used to describe an event or situation rather than as dependent on the event or situation itself.

A similar argument shows that probabilities intuitively depend not only on the properties used to describe the particular situation but on what is regarded as a repetition of these properties – on the spacing parameter in effect. Suppose in the above case we allow 100 men from the same town to count as 100 repetitions. This is covertly introducing the extra describing property, 'Englishman from town X', and this property will intuitively affect the probability. What happens in practice is that the spacing condition is carefully and subtly specified so that one obtains a 'random sample', i.e. so that the axiom of independent repetitions is satisfied.

Lastly, the probabilities of particular events depend not only on the set of repeatable conditions which these particular events instantiate but also on what outcomes of the conditions they are taken to be. I can quote a classic example (usually attributed to Laplace, but in fact invented by d'Alembert) to illustrate this point. Suppose we have apparatus which will deliver any letter of the alphabet with equal probability. Suppose we obtain 14 successive letters from this apparatus and these spell out CONSTANTINOPLE. On the one hand we would feel inclined to say that this was a very unlikely happening but on the other the probability of this combination of letters is surely the same as any other, viz. $(1/26)^{14}$. The explanation of this riddle is given by von Mises (1928, pp. 19–20). In this case we can choose our space of outcomes Ω in two ways. Either we could take Ω to comprise the $(26)^{14}$ different combinations of 14 letters (Ω_1) or to comprise the two attributes 'meaningless' or 'nonsensical' (Ω_2). Considered as an outcome in Ω_1 the result is not more improbable than any other result. Considered as an outcome in Ω_2 the result is very very much less probable than the other possible outcome. When we speak of the 'outcomes' of a certain set of conditions S_1 we are really referring to a set of conditions S_2 s.t. if S_1 is satisfied, one and only one member of the set S_2 will be satisfied. However, for fixed S_1, S_2 can be chosen in a variety of ways.

These considerations suggest the following account of probability which is the best I can do by way of a philosophical analysis of the concept. In certain circumstances we find that if a certain set of repeatable conditions S_1 is instantiated a certain further condition is also instantiated. In this case we obtain a deterministic law. In other circumstances given a certain S_1 we cannot predict that a further condition will be instantiated, but only that one of a certain set of conditions S_2 will occur. On the other hand we can give a greater weight to certain of the conditions in S_2 than to others. We postulate that these weights satisfy certain laws and stipulate how this calculus of weights is to be used in practice. From these assumptions we can deduce that those conditions with the greater weight will in certain sequences of repetitions occur more often. This enables us to test the calculus and to use it in applications. These situations are those which involve probabilistic rather than deterministic laws. Finally we can introduce probabilities for single instances of the set S_1 of conditions. These singular probabilities are, however, only validly defined w.r.t. the given set of repeatable conditions and the given set of possible outcomes under which the single event in question is subsumed. We have to describe a particular event or situation in terms of a particular conceptual scheme before we can assign probabilities to certain features of it.

We now argue that this 'conditionality' is not a serious restriction. Consider the example of building the refinery which was our paradigm case of the introduction of probabilities for single events. Before the matter could be handled mathematically at all the situation would have to be subsumed under a general and abstract description. Now according to our assumption there cannot for contingent reasons be two instantiations of this description. On the other hand certain parts of the description might be repeated more than once, and the interaction between the various conditions might be of a kind that had been studied elsewhere. In this way it might be possible to postulate a reasonable probabilistic theory of the situation and so obtain valuable results.

(vii) An answer to some objections of Ayer's

For this reason I do not consider that the 'conditionality' of singular probabilities constitutes a serious problem. This is not the view of Ayer, who in his second note on probability (1963, pp. 198–208) argues that this difficulty constitutes a fatal or at least highly serious objection to the frequency theory. Although his argument is in terms of the frequency theory, it applies equally to the approach advocated here. Ayer begins by considering the probability that he (Ayer) will live to be 80. He points out, just as I have done earlier, that this probability depends on whether he is considered as a member of the class of organisms in general, the class of all human beings, the class of male Europeans.... He then asks (1963, p. 200): 'What reason can there be, within the terms of the frequency theory of probability, for basing the estimate of my chances of longevity on the ratio obtaining in any one of these classes rather than any other?' My answer is that there is indeed no reason *within the frequency theory* (or within our own theory for that matter), but this is no argument against the frequency theory because it is not its business to supply such reasons. The reasons for preferring one class to another are given by considerations external to the frequency theory. The nature of these 'external reasons' can only be seen by analysing particular practical cases.

The obvious instance to choose here is the application of probability theory to the insurance business. The 'logic of the insurance business' can be described roughly thus. An insurance company wants to calculate a reasonable premium, say for 40-year-old Englishmen who insure themselves against dying before reaching the age of 41. The outlay of the company is determined by the relative frequency of its clients in this category who do actually die before they are 41. Let us call this frequency f. The clients are considered as a long sequence of independent repetitions of a certain set of conditions $\mathfrak{S}_s$ and it is assumed that relative to $\mathfrak{S}_s$ we have prob(death) $= p$ say. Then from the binomial distribution we obtain that there is a very high probability that $p \doteqdot f$. The value of p is estimated from other data, which is also considered as a sequence of repetitions of $\mathfrak{S}_s$; and hence an estimate of f is obtained. Our problem here is how do we choose $\mathfrak{S}_s$? Under how many properties do we classify the clients? The answer is determined by two factors. On the one hand we want to have enough data to be able to estimate p accurately. On the other we want to be able to charge a premium which is competitive with other companies. The first factor will tend towards choosing for $\mathfrak{S}_s$ a broad and hence often instantiated set of conditions. The second will have the opposite effect. If we make the conditions $\mathfrak{S}_s$ broad we may include certain special cases (men who for some reason are at death's door) which will have the effect of increasing the value of p considerably. By narrowing the conditions $\mathfrak{S}_s$ so as to exclude these cases the value of p will be reduced and so the premium charged to the majority of the clients can be reduced without fear that the company will go bankrupt. The special features of the situation will decide how these two factors are balanced in the final choice of $\mathfrak{S}_s$. Of course what I have described is a considerable simplification of the insurance problems which occur in practice – but I hope, nonetheless, it is adequate for illustrating the logic of this kind of application.

PART THREE

A Falsifying Rule for Probability Statements

CHAPTER EIGHT

The Falsification Problem for Probability Statements

(i) Statement of the problem

In deriving the empirical laws of probability from the axioms of the theory, we had to make use of what I called a 'methodological rule for neglecting small probabilities under certain circumstances'. Apart from an analogy with Newtonian mechanics, I have so far offered no justification for such a rule, and it is time now to look into the matter more closely. The problem is perhaps best approached by way of a difficulty in Popper's account of scientific method, which was clearly explained by Popper himself as follows (1934, p. 146):

> The relations between probability and experience are also in need of clarification. In investigating this problem we shall discover what will at first seem an almost insuperable objection to my methodological views. For although probability statements play such a vitally important rôle in empirical science, they turn out to be in principle *impervious* to *strict falsification*. Yet this very stumbling block will become a touchstone upon which to try my theory, in order to find out what it is worth.

It is easy to see why probability statements cannot be falsified. Suppose we are dealing with a random sequence of heads and tails and postulate that $\mathrm{prob\,(heads)} = p$. Then we have

$$\mathrm{prob}\,(m) = {}^{n}C_{m}\, p^{m}\, q^{n-m}$$

where $q = 1 - p$ and $\mathrm{prob}\,(m)$ is the probability of getting m heads in the first n tosses. So however long we toss the coin (i.e. however big n is), and whatever number of heads (i.e. whatever value of m) we observe, our result will always have a finite probability. It will not be strictly ruled out by our assumptions. In other words these assumptions are 'in principle *impervious to strict falsification*'.

This result can even be strengthened slightly in the following way. Let us select any ordered n-tuples of heads and tails, and a probability arbitrarily close to 1 (say $1 - \epsilon$). We can then find N such that in the first N elements of a random sequence of heads and tails with prob (heads) $= p$, there is a probability of $1 - \epsilon$ of finding our particular n-tuple. Put less formally we can say that in a sufficiently long series of tosses any particular n-tuple (say n heads) is almost certain to appear. If we get this n-tuple as the result of our experiment why should we not say 'this result has just happened to appear now rather than later'?

(ii) Popper's proposed solution

What conclusion are we to draw from this? At first it would seem that we have a clear counter-example to Popper's demarcation criterion. After all it must be admitted that probabilistic theories appear in much reputable science. We have just argued that they are not falsifiable. We must then conclude apparently that not all scientific theories are falsifiable. Popper however has an answer. It is this. Although strictly speaking probability statements are not falsifiable, they can nonetheless be used as falsifiable statements, and in fact they are so used by scientists. He puts the matter thus (1934, p. 191): '...a physicist is usually quite well able to decide whether he may for the time being accept some particular probability hypothesis as "empirically confirmed" or whether he ought to reject it as "practically falsified"....'

Another possible solution would be to say that although probability statements cannot be strictly falsified they can nonetheless be corroborated by evidence to a greater or lesser extent. We could thus have rational reasons for preferring one statistical hypothesis to another. Even without introducing the notion of corroboration we might still be able to devise rational policies for deciding between different statistical hypotheses on the basis of evidence. I shall not consider such a confirmation theoretic approach in detail. I can however point out that the general arguments in favour of Popper's falsifiability criterion tell against such an approach. In effect if a theory is to tell us anything interesting about the world, it cannot surely be compatible with every possible state of affairs. But if some state is ruled out, that state becomes a potential falsifier of the theory and the theory becomes falsifiable. Popper buttresses this general argument by pointing out that if we use probability hypotheses in a non-falsifiable fashion we can thereby explain anything at all. For example, we can explain Newton's laws on the assumption that all the particles in the universe are randomly scattered with regard to position and velocity. Under this hypothesis of random scatter there is still a finite probability of Newton's laws being obeyed 'by accident' over a long period. Finally (employing the device of the previous paragraph) we can say that if the universe persists for a sufficiently long time it becomes 'almost certain' that this accidental obedience to Newton's laws will occur. Popper, after giving this example, concludes (1934, p. 198): 'It seems clear to me that speculations of this kind are "metaphysical", and that they are without significance for science.'

In fact, however, Popper later discovered arguments similar to his example in the writings of certain eminent scientists. (He gives one example in his footnote 1 to p. 198.) Strictly speaking this disproves the assertion that scientists always use probability statements as falsifiable statements. But this ought anyway to be reformulated so as to become *normative* as well as factual. We might say: 'Probability statements *are* usually used by scientists in a falsifiable fashion, and *ought* for general reasons always be so used.'

(iii) The problem of producing an F.R.P.S.

This defence of Popper's raises a problem which I propose to call *'the problem of producing an F.R.P.S.* (or Falsifying Rule for Probability Statements)'. In other words, assuming that Popper is right and scientists do use probability statements as falsifiable statements, the problem is to give an account of *how* they do so; to give explicitly the methodological rules which implicitly guide their handling of probability. There are at least two adequacy requirements on any proposed F.R.P.S. First of all it must correspond in some sense to the way that scientists do in fact behave. Of course we can again make the point that such a rule is normative as well as factual. A philosopher is surely entitled to criticize certain portions of science for not satisfying what seem to him to be desirable methodological principles. But of course if he finds that according to his methodology huge portions of science become vicious, there is little doubt what he ought to reject. Secondly I think it is necessary that there should be general arguments in favour of accepting the F.R.P.S. If the best we can produce is entirely arbitrary, this would suggest that probability ought to be dealt with in a different way altogether. The fact is that we could make a number of arbitrary stipulations which would render probability statements falsifiable, but if any one of these stipulations was either (i) *unrealistic* (in the sense of not corresponding to scientific practice) or (ii) *unreasonable* (in the sense of not having any general considerations to support it), it could not be taken as constituting an adequate F.R.P.S.

I shall attempt to formulate an F.R.P.S. and to examine its relation to statistical practice in the next chapter. Then in Chapter 10 I will examine what general considerations can be advanced in favour of the rule. Before starting on this projected solution, however, it will be interesting to add some notes about the history of the problem.

(iv) Some notes on the history of the problem
Popper was the first to formulate the 'falsifiability' criterion to demarcate science from metaphysics, and correspondingly the first to give a clear statement of the falsification problem for probability statements. However, as we might expect, there was some awareness of the difficulty among writers on probability before Popper. The most remarkable anticipation of Popper's ideas occurs in d'Alembert. Todhunter describes d'Alembert's views as follows (1865, p. 262):

> D'Alembert says that we must distinguish between what is *metaphysically* possible, and what is *physically* possible. In the first class are included all those things of which the existence is not absurd; in the second class are included only those things of which the existence is not too extraordinary to occur in the ordinary course of events. It is *metaphysically* possible to throw two sixes with two dice a hundred times

running; but it is *physically* impossible, because it never has happened and never will happen.

What is surprising here is the use of the same terms 'metaphysical' and 'physical', and in a sense not very different from Popper's. Although no set of results obtained by throwing two dice a hundred times is strictly ruled out, we agree that some such sets are so improbable as to be ruled out physically, i.e. from a practical point of view. The question now arises: how small a probability must an event have for us to regard it as 'physically impossible'? Or, to put it another way, when can we take a small probability as being equivalent to zero? D'Alembert regards this as one of the most important problems connected with probability theory, though, at the same time, he seems to consider it as perhaps insoluble. He in fact says (quoted from Todhunter 1865, p. 265): '...in order to arrive at a satisfactory theory of probability, it would be necessary to resolve several problems which are perhaps insoluble'; the second of these is 'to determine when a probability can be regarded as zero'.

Buffon took up this problem and suggested that a probability of 10^{-4} or less could be regarded as zero (cf. Todhunter 1865, p. 344). He justified this figure as follows. From mortality tables he discovered that the chance of a man aged 56 dying in the course of a day was approximately 10^{-4}. He then argued that such a man considers this probability as equivalent, from a practical point of view, to zero. Buffon's reasoning seems a trifle quaint; but it is genuinely difficult to see how we could determine a value which is not wholly arbitrary.

The approach of d'Alembert and Buffon, translated into our own conceptual scheme, suggests the following falsifying rule. Suppose that, given a statistical hypothesis H, it follows that a certain event ω has a probability p where $p \leqslant p_0$, p_0 being some suitably small value such as 10^{-4}. Then if ω is observed to occur we must regard H as falsified from a practical point of view.[1]

[1] i.e. we must treat H as we would treat a deterministic hypothesis which had been falsified in a strict or logical sense. This is not of course equivalent to rejecting H. For example, Newton's theory was falsified by the motion of Mercury's perihelion many years before it was rejected. There is no reason in principle why something similar should not happen in the statistical case.

Let us call this first attempt at a falsifying rule R.1. I think it is the rule which most naturally suggests itself, and as we shall see, it has been adopted not only by d'Alembert and Buffon but by many later statisticians. At all events we will begin our own attempt to formulate a falsifying rule by discussing, and rejecting, it.

CHAPTER NINE

Formulation of a Falsifying Rule

(i) First version of the rule

Rather than give my suggested version of an F.R.P.S. straight away, I will consider some preliminary versions which might more naturally suggest themselves. By criticizing and improving these I will proceed to the rule which will finally be advocated. As indicated at the end of the last chapter I will begin with the rule of d'Alembert and Buffon (R.1) which stated roughly that we will regard a hypothesis H as falsified if the observed event has a low probability given H. Against this rule a simple but fatal objection can be raised. I have in fact already mentioned this point. Suppose we are tossing a coin and have the hypothesis that the events are independent with prob (heads) $= \frac{1}{2}$. Suppose we have a million tosses and observe exactly the same number of heads and tails. The probability of this event is of course

$$^{1,000,000}C_{500,000} \left(\tfrac{1}{2}\right)^{1,000,000} = 0 \cdot 0008 \qquad \text{(to 1 sig. fig.).}$$

Now $0 \cdot 0008$ is a very small probability. Therefore applying R.1 with, say, $p_0 = 10^{-3}*$ we would be forced to conclude that the observation falsified the hypothesis. In practice however one would draw the opposite conclusion, namely that the evidence was almost the best possible in support of the hypothesis. Worse still, the probability of any other proportion of heads and tails is less than $0 \cdot 0008$, and so applying R.1 we would have to regard H as falsified whatever happened.

It is not difficult to see the error involved here. What is important is not the actual probability of a particular outcome, but the relation between its probability and the probabilities of other possible outcomes. Thus the actual value of $0 \cdot 0008$ is low,

---

* This is rather higher than Buffon's suggested value of 10^{-4}, but the latter seems unduly small.

but it is nonetheless very much greater than the probability of getting 1,000,000 heads. This suggests that we should introduce a measure of the probability of a given event relative to the other possible events, and only regard a hypothesis H as falsified if, relative to H, the observed event has a low value on this measure. In order to do this we must first formulate the problem in a manner which is rather more precise mathematically.

(ii) Second version of the rule

We will suppose that from our statistical hypothesis H we deduce that a certain $1 - D$ random variable ξ has a distribution D. We will further require that D is either a discrete or continuous distribution. In general we shall denote the range of ξ, i.e. the possible values it can take, by $R(\xi)$. In the discrete case $R(\xi)$ will consist of a finite or denumerable set (x_i) $(i = 1, 2, \ldots)$ with probabilities $p(\xi = x_i) = p(x_i)$. In the continuous case $R(\xi)$ will be some subset of the real line, and we will suppose ξ has a probability density function $f(x)$. For reasons which will become clear later, we will require that the maximum value of $f(x)$ is finite. Our problem is to decide which possible values of ξ (i.e. which members of $R(\xi)$) should, if observed, lead us to reject the hypothesis that ξ has the distribution D and hence the original statistical hypothesis H.

This formulation obviously corresponds very closely to the usual case of statistical testing. There we have a random sample $\xi_1 \ldots \xi_n$ drawn from a population for which we have postulated a hypothetical probability distribution. We calculate a statistic $\eta(\xi_1, \ldots, \xi_n)$ which is a function of $\xi_1, \ldots, \xi_n$. Since the ξ_i are r.v.'s, η is a $1 - D$ r.v. and its distribution D can be calculated from the population distribution. In practice D is always a discrete or a continuous distribution. The problem is then whether the observed value of η is compatible with its hypothetical distribution D. Sequential testing procedures can also be fitted into this framework. Finally we can remark that the restriction to $1 - D$ r.v.'s is not essential. However, it greatly simplifies the problem and really involves no great loss of generality. Suppose we have a number of r.v.'s $\xi, \eta, \ldots$ with joint distribution D'. We can, as already indicated, consider a $1 - D$ statistic $\zeta(\xi, \eta, \ldots)$ whose distribution D is determined from D'.

We now introduce the concept of the *relative likelihood* $l(x)$

Formulation of a Falsifying Rule 169

of a possible result $x \in R(\xi)$. This is defined as follows. Let $p_{\max}$ be the maximum value of the $p(x_i)$ in the discrete case, and $f_{\max}$ be the maximum value of $f(x)$ in the continuous case. We then set

$$l(x) = p(x)/p_{\max} \quad \text{in the discrete case}$$
$$= f(x)/f_{\max} \quad \text{in the continuous case.}$$

$l(x)$ gives a measure of the probability of an observed event in relation to the other possible events. We can now give our second version of an F.R.P.S. which depends on this measure.

R.2. If we observe a value x of ξ with $l(x) < l_0$ where l_0 is some suitably small value, then we must regard H as falsified.

This rule is a considerable improvement on the previous one. To begin with it avoids the simple coin-tossing counter-example. The result 500,000 heads has an l value of 1 – the greatest possible. Whereas the l value of, say, 1,000,000 heads is very low. An observation of the latter, but not of the former, would lead to a rejection as we intuitively require. Nonetheless R.2 in its turn falls to a rather more sophisticated counter-example.

Suppose the r.v. ξ can take any of the integral values $0, 1, 2, \ldots,$ 9,900. Suppose its distribution D is given by

$$p(\xi = 0) = 0 \cdot 01$$
$$p(\xi = i) = 10^{-4} \quad (i = 1, 2, \ldots, 9,900)$$

for $i = 1, 2, \ldots 9,900$ we have $l(i) = 10^{-2}$. Assuming that this is small enough to give a falsification, we have in accordance with R.2 that H should be rejected if we get a value i with $1 \leqslant i \leqslant 9,900$, and accepted only if we obtain $\xi = 0$. But then if H is true there is a probability of $0 \cdot 99$ or 99 per cent that it will be regarded as experimentally falsified and of only 1 per cent that it will be regarded as experimentally corroborated. This is clearly unsatisfactory.

(iii) Third version of the rule

Once again it is not difficult to see the root of the trouble here. We must require not only that the observed result x should have a small relative likelihood but that it should be *untypical* in this. If nearly all the possible results had low relative likelihoods, as in the counter-example just given, then we would actually expect to get a result with low relative likelihood and would not regard it as falsifying the hypothesis. We must therefore in each case consider the class of possible results x with $l(x) < l_0$, and require that this class has a low probability given H. In this way we are led to the familiar statistical concept of a critical region. Consider the range $R(\xi)$ of ξ. We can look on the problem as that of finding a subset C of $R(\xi)$ s.t. if the observed value of ξ (x say) is in C, we regard H as falsified. C is then called a 'critical region' or a 'falsification class'. Let us call the probability that we get a result $x \in C$, $k(C)$ (or simply k). We must then require that k is small – say $k < k_0$. In addition we have to formulate the considerations of relative likelihood already introduced. Let us define the relative likelihood of any arbitrary subset B of $R(\xi)$ by the formula

$$l(B) = \operatorname*{def\,max}_{x \in B} l(x).$$

The relative likelihood of the critical region, $l(C)$, we shall usually denote simply by l. Our previous discussion now leads to the requirement that $l < l_0$ for some suitably small value l_0. Finally let us denote the set $(R(\xi) - C)$ by A. A is the 'acceptance region' of the test. It seems reasonable to require that the relative likelihood of any of the results x in A should be greater than the relative likelihood of any results in C. Putting all this together we reach our third version of an F.R.P.S.

R.3. Suppose that the range $R(\xi)$ of ξ can be partitioned into two disjoint subsets C and A with $C \cup A = R(\xi)$ and s.t.

(i) prob $(\xi \in C) = k < k_0$;

(ii) $l(C) = l < l_0$, where k_0 and l_0 are suitably small constants; and

(iii) $l(x) > l$ for all $x \in A$.

We shall then regard H as falsified if the observed value of ξ, x say, lies in C.

This rule is very near our final version, but it is not satisfactory because the requirement (iii) turns out to be too weak. To see this, consider the following counter-example which is a modification of our previous one. Suppose a r.v. ξ can take on the integral values 0, 1, 2,..., 9,940. Suppose further ξ has the distribution D defined by

$$p(\xi = 0) = 0 \cdot 01$$
$$p(\xi = i) = 10^{-4} \qquad (i = 1, 2, \ldots, 9{,}540)$$
$$p(\xi = i) = 9 \times 10^{-5} \qquad (i = 9{,}541, \ldots, 9{,}940).$$

Suppose we now set

$$A = (0, 1, 2, \ldots, 9{,}540)$$
$$C = (9{,}541, \ldots, 9{,}940)$$

Then

(i) $k(C) = 0 \cdot 036 < 0 \cdot 05$ (the 5 per cent level);

(ii) $l(C) = 9 \times 10^{-3}$ (presumably a sufficiently low value); and

(iii) $l(x) > l$ for $x \in A$.

Thus the partition (A, C), as defined above, satisfies the requirements of R.3. It does not, however, strike me as satisfactory from an intuitive point of view. My reason is that the results of the rejection class have relative likelihoods which are not significantly less than most of the results in the acceptance class A. It does not seem sensible to reject the hypothesis if we get a result with relative likelihood (9×10^{-3}) when nearly all the results have a relative likelihood (10^{-2}) which is only marginally greater. What makes this case unsatisfactory is that the value $p_{\max}$ is in no way typical of the majority of probabilities of results in the acceptance class. In fact it is very untypical. Consequently I suggest that condition (iii) in R.3 be strengthened to state that $p_{\max}$ is in some sense representative of the probabilities $p(x)$ for $x \in A$. This is obviously a very vague requirement but we will not, at this stage, attempt to make it more precise. If we add this qualification to R.3 we obtain the F.R.P.S. which will in fact be advocated in what follows. We will now try to state it in a reasonably precise fashion.

(iv) Final version of the rule

It will prove convenient to introduce a new concept in terms of which the F.R.P.S. can be simply stated. Our procedure is to partition $R(\xi)$ into two disjoint sets C and A which satisfy certain conditions. Now it will obviously not always be possible to do this for a random variable with any arbitrary distribution. Where it is possible we shall speak of a random variable with a *falsifiable* distribution. More precisely this can be defined as follows. We shall say that a $1 - D$ r.v. ξ has a *falsifiable* distribution if it is possible to partition $R(\xi)$ into disjoint sets A and C with $A \cup C = R(\xi)$ where

(i) prob $(\xi \in C) = k < k_0$

(ii) $l(C) = l < l_0$ where k_0, l_0 are suitably small constants, and

(iii) the value $p_{\max}$ (or $f_{\max}$ in the continuous case) is in some sense representative of the probabilities (resp. probability densities) of points $x \in A$.

If a r.v. ξ has a falsifiable distribution D, we shall call a set C a 'critical region' or 'falsification class' associated with D, and a set A an 'acceptance region' associated with D. In terms of these concepts we can now formulate our falsifying rule as follows.

F.R.P.S. If from a statistical hypothesis H we can deduce that a r.v. ξ has a falsifiable distribution D, and if C is a critical region associated with D, then if a value x of ξ with $x \in C$ is observed we regard H as falsified. We will now make a number of comments about this F.R.P.S.

First of all, when we test H by means of ξ we can be said to be predicting $\xi \in A$, and are regarding our prediction as falsified if $\xi \notin A$. As I remarked earlier, in the deterministic case 'predicting' and 'explaining' are in a sense symmetrical. This symmetry can be restored in the statistical case if we introduce the notion of 'deducibility in accordance with the falsifying rule' or 'f-deducibility' for short. This can be made precise as follows. Suppose that from a statistical hypothesis H we deduce (logically) that a certain r.v. ξ has a falsifiable distribution D with associated acceptance region A. We shall then say that $\xi \in A$ is f-deducible from H. We will say that a certain frequency phenomenon is explained by a statistical hypothesis H if the frequency phenomenon can be described by a statement s of the form '$\xi \in A$' where s can be f-deduced from H. If we extend the concept of deducibility to include f-deducibility, we thereby preserve the so-called 'deductive model of explanation'. The symmetry between 'explaining' and 'predicting' is again clear. In both cases we f-deduce that $\xi \in A$. If $\xi \in A$ has already been established experimentally, we say that it has been explained. If it has not been so established, we have a prediction. If ξ is then in fact observed to be in A, the prediction is verified and the hypothesis corroborated. If, however, we find $\xi \notin A$, the prediction and consequently the hypothesis is falsified. These points are straightforward, but nonetheless important. Most statisticians who have considered this problem have concentrated their attention on testing statistical hypotheses rather than on using such hypotheses to explain frequency phenomena. As a result there are certain theories of testing which allow statistical hypotheses to be tested, but do not allow us to use such a hypothesis to explain an observed phenomenon. Such theories cannot in my view be regarded as satisfactory.

It is often said that statistical laws tell us nothing about the particular case, but only about what happens in a large number of similar cases. This is not a very satisfactory formulation since a large number of similar trials can be considered as a single combined trial, and statistical laws do tell us what happens on trials *of this form*. The notion of a falsifiable distribution helps to make clear what truth there is in this remark. Suppose we have a random variable ξ. We can only make prediction about the observed value of ξ if ξ has a falsifiable distribution, and in that case we can f-deduce and hence predict that $\xi \in A$, where A is an acceptance region. Let ν be a random variable denoting the number of heads observed in a coin-tossing experiment. Suppose we toss the coin just once, then ν does not have a falsifiable distribution. In fact ν has the uniform distribution $\text{prob}(\nu = 0) = \text{prob}(\nu = 1) = \frac{1}{2}$ which is almost the paradigm case of a non-falsifiable distribution. Suppose we toss the coin a large number n of times, then ν will have a falsifiable distribution and we will be able to make predictions about how many heads will be observed.

(v) Application of the rule to an important special case
Let us now see how the F.R.P.S. applies to a simple but very important case. Suppose ξ has a continuous, unimodal 'bell-shaped' distribution of approximately normal form, as shown in diagram 1 below.

Obviously the F.R.P.S. will instruct us to choose an acceptance interval of the form $[a, b]$ and a corresponding critical region $(-\infty, a) \cup (b, +\infty)$; in other words, to divide the range of ξ into a 'head' and two 'tails', and to reject H only if we obtain an observed value of ξ in the tails. We could term this procedure: 'cutting off the tails of the distribution'. We have

$$k = \int_{-\infty}^{a} f(x)\,dx + \int_{b}^{\infty} f(x)\,dx$$

and
$$l = \max\left(f(a)/f_{\max}, f(b)/f_{\max}\right).$$

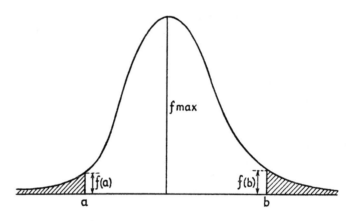

Diagram 1. The F.R.P.S. illustrated in a simple case.

Let us now examine the significance of the condition that f_{max} should, in some sense, 'be representative of the values of $f(x)$ in $[a,b]$'. Another way of putting this is as follows. We require that $f(x)$ should have a low value in the 'tails' of the distribution, but once we enter the 'head' or 'acceptance region' $[a,b]$, we would like $f(x)$ to rise to its maximum value f_{max} as quickly as possible. So we require that f should increase swiftly once inside the region $[a,b]$; but how swiftly? I believe that, at the cost admittedly of some arbitrariness, we can give a precise answer to this question. This criterion then enables us to say definitely whether a distribution of the continuous, unimodal form is falsifiable or not, and if it is falsifiable to divide it into a 'head' and 'tails'. This procedure is of course very useful in comparing the recommendation of our F.R.P.S. with the methods actually used in statistical practice. However, I will not here give the details of this further attempt at preciseness. Rather I will assume that the qualitative considerations given to date enable us, roughly at least, to divide continuous, unimodal distributions into those which are falsifiable and those which are not and to define the boundaries of the related acceptance regions.

I must now say a few words about the interpretation of the constants k and l. In the usual approach the constant k (the *size* of the critical region) is wholly arbitrary. We have to begin by selecting a suitable value, and if we then want to change to another value we can correspondingly alter our falsification class without difficulty. On my own approach, it is not such an easy matter to alter k. Suppose we want to test a hypothesis H by observing one value of a random variable ξ. Suppose further that ξ has a continuous, unimodal distribution. For the test to be possible at all we must first have that ξ has a falsifiable distribution. Suppose this is so. Then, as we just remarked, we can use geometrical considerations to divide the range of ξ into a head and tails. But now k is determined as the probability of getting a result in the tails. We therefore have that, given the distribution, the values of k (and indeed l) are determined. If we want to change the value of k, say, we cannot, as on the standard approach, do so by a stroke of the pen. We have to find a new random variable η s.t., given the hypothesis H, η has a falsifiable distribution with a critical region whose k value is different. This is not likely to be an easy task.

k and l can be taken as measuring the difference between the statistical test in question and a corresponding deterministic test. In the deterministic case we predict that, given H, we will certainly have $x \in I$ where x is some observation and I is an interval. In the statistical case we make a similar prediction, but only with some k- and l-values. The deterministic case can be taken as a limiting case of this with $k = l = 0$, and in general the values of k and l give a guide as to how much the statistical differs from the deterministic case. Of course we have still got to fix the crucial values k_0 and l_0, i.e. the degree of divergence from the deterministic case which we consider allowable. However, the preceding considerations about the geometry of distributions give us some guide on this too.

The argument is this. It will be granted that the normal distribution plays a fundamental role in statistics – for example in the theory of errors. Now the normal distribution is evidently of falsifiable form. If we divide its range into an acceptance region and a critical region, using the rough geometrical considerations explained above, we obtain $k \doteq 3$ per cent, $l \doteq 10$ per cent. As we will naturally expect the normal distribution to have better properties than can be hoped for in general, this suggests giving k_0 and l_0 rather higher values than the above ones. We are thus led to the suggestion $k_0 \doteq 5$ per cent, $l_0 \doteq 15$ per cent.

This provides some kind of justification, albeit a weak one, for the 5 per cent value customarily used by statisticians.

We are now in a position to examine whether or not our proposed rule agrees with statistical practice. I think it is clear that the rule accords well with the standard statistical tests. Consider the χ^2-, t-, and F-distributions. For sufficiently high degrees of freedom, these are all of the continuous, unimodal form already discussed. Using our rough geometrical considerations, we can divide their ranges into a head and tails, and it turns out that the corresponding k- and l-values are of the right order. In this way the χ^2-, t-, and F-tests are justified from our point of view.

However, at the same time a certain difference between the present approach and standard practice makes itself manifest. Sometimes a one-tailed version of say the t-test is used. Now this is certainly illegitimate from our point of view. Suppose the tail used as the critical region C in the one-tailed test is the right-hand one. Then we can certainly find points x in the acceptance region s.t. $l(x) < l(C)$. Points in the left-hand tail of the curve can be chosen for this. This contradicts even the weaker preliminary version R.3 of our falsifying rule, and thus evidently the final stronger version.

One-tailed tests are used because they are suggested by the Neyman–Pearson theory. So a difference between the present approach and the Neyman–Pearson theory has come to light. I will now examine the relations between the two accounts in rather more detail.

(vi) Relation of the present approach to the Neyman–Pearson theory

Our F.R.P.S. forced us in the continuous unimodal case just discussed to take for our falsification class the 'tails' of the distribution. Only if we chose the tails could we obtain a low l-value as required. Consider however a falsifying rule of the following form (which we shall call R.4).

R.4. Suppose the range $R(\xi)$ of a random variable ξ with postulated distribution D can be partitioned into two disjoint sets A and C with $A \cup C = R(\xi)$, and suppose $\text{prob}(C) = k < k_0$ where k_0 is some suitable constant. Then if we observe $\xi \in C$, the hypothesis that ξ has distribution D must be taken as falsified. With such a rule we would not be restricted to the 'tails' of the distribution, but could for example choose for C any interval $[c,d]$ satisfying $\int_c^d f(x)dx = k < k_0$.

Some 'counter-intuitive' choices of this type are illustrated in diagram 2 below.

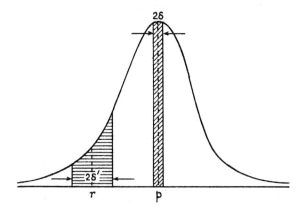

Diagram 2. Showing 'counter-intuitive' choices of falsification class.

Suppose again we have the experiment of tossing a biased coin for which prob(heads) $= p$. Consider the frequency ratio m/n of heads in a large number n of tosses. For large n this random variable is approximately normally distributed with mean $=$ mode $= p$ as shown in diagram 2. Now if we adopted R.4 we could choose as rejection class C an interval $(p - \delta, p + \delta)$ where δ is sufficiently small to ensure that $\int_{p-\delta}^{p+\delta} f(x)dx = k < k_0$. Correspondingly we would reject the hypothesis if m/n fell extremely near to p. This is the same intuitive contradiction which led us to reject R.1. We could moreover adopt any arbitrary interval $(r - \delta', r + \delta')$ say where δ' is chosen so that $\int_{r-\delta'}^{r+\delta'} f(x)dx = k < k_0$. Evidently such a procedure is completely arbitrary.

The problem presented here is closely connected with the Neyman–Pearson theory of testing hypotheses. Roughly speaking Neyman and Pearson begin by adopting a falsifying rule of the form R.4. They then note the paradoxical choices of rejection class as in diagram 2. Then in order to escape the difficulty, they add a new principle to the effect that we cannot test a given hypothesis in isolation, but only against a set of alternative hypotheses. Developing from this position they show how the counter-intuitive falsification classes can be ruled out. Our own procedure is somewhat different. We eliminate these counter-intuitive falsification classes by requiring a low l- as well as a low k-value. We say in effect that the falsification class must have a low relative likelihood as well as a low probability. We thus solve the problem without appeal to a principle of alternative hypotheses, and it consequently becomes possible, on our position, to test a statistical hypothesis 'in isolation', i.e. without considering well-defined alternative hypotheses.

A second difference from the Neyman–Pearson approach lies in our notion of a 'falsifiable distribution'. On our account it is only random variables ξ with *certain* distributions D, whose ranges $R(\xi)$ can be divided into an acceptance region A and a critical region C. Suppose however that we adopted a falsifying rule of the form R.4. Then any random variable ξ with *any* distribution D could have its range $R(\xi)$ partitioned into an A and a C. There would be no distinction between 'falsifiable' and 'non-falsifiable' distributions.

The difference is most vividly shown in the case of the rectangular distribution. Suppose ξ has a frequency function $f(x)$ defined by

$$f(x) = 1 \qquad \text{for } -\tfrac{1}{2} \leqslant x < \tfrac{1}{2}$$
$$= 0 \qquad \text{otherwise.}$$

Adopting R.4 we could choose for our critical region C any interval (a,b) where $-\tfrac{1}{2} \leqslant a < b < \tfrac{1}{2}$ and $b - a = k$, where k is some suitable constant $< k_0$. I shall show in Chapter 11 that on the Neyman–Pearson theory a number of critical regions C of the form (a,b) are possible depending on which 'class of alternative hypotheses' we adopt. On our own view, however, no critical region of the form (a,b) is allowable. For any such region we have $l = 1$, i.e. its maximum possible value. Hence we certainly shall not have $l < l_0$ whatever 'crucial' value l_0 is chosen. I would claim that the non-existence of a falsification class here is in accordance with intuition. Suppose we have, say, $-\tfrac{1}{2} < a < b < a' < b' < \tfrac{1}{2}$, and $b' - a' = b - a = k$. If we now adopt (a,b) as our critical region, we will reject the hypothesis if the observed value $x \in (a,b)$ and not if $x \in (a',b')$. Yet as far as the hypothesis under test goes, the two regions (a,b) and (a',b') are exactly symmetrical. It seems wholly arbitrary to adopt one and not the other as a critical region. The only solution is to say that this particular distribution is not one of the falsifiable kind.

Thirdly we have the point already mentioned, that, on the Neyman–Pearson theory, we can sometimes have one-tailed tests where, from our point of view, only two-tailed tests are valid. These points show that there is a clear difference between the view presented here and the Neyman–Pearson theory. Now the Neyman–Pearson theory is the generally accepted account of testing statistical hypotheses. Thus if the present view is to gain any credence, we must give reasons why the Neyman–Pearson theory should be abandoned. This task will be attempted in Chapter 11. In the next chapter I will examine what general considerations can be advanced in favour of my proposed rule.

CHAPTER TEN

Evaluation of the Falsifying Rule

(i) Braithwaite's theory of probability: statement

A convenient introduction to the problems of this chapter will
be provided by an examination of Braithwaite's theory of
probability. This is a doubly desirable task since our own theory
is very similar to Braithwaite's and indeed was greatly influenced
by the latter. It is therefore most necessary to compare the two
positions and to explain how far they are in agreement and
where and why they differ. Braithwaite's theory is expounded
in *Scientific Explanation*, Chapters V–VII. I will now attempt
a brief outline of his position.

Braithwaite takes the fundamental problem of a philosophical
theory of probability to be that of explaining how probability
statements acquire meaning. Indeed the title of one of his
chapters on probability is (1953, Ch. VI) 'The *meaning* of
probability statements within a scientific system' (my italics).
Earlier in *Scientific Explanation* he propounds a general theory
of how the theoretical terms of science acquire meaning and it
is in terms of this theory that he deals with the case of probability.
In this theory two components of the meaning of a theoretical
term like 'electron' are distinguished. These could be called its
'logical' and its 'empirical' meaning. To begin with, the symbol
'*e*' appears in a large number of abstract mathematical theories
or calculi. These calculi provide rules relating the symbol *e* to
other symbols and explaining how it can be manipulated. Such
rules of the calculus give *e* its *logical* meaning. However we also
derive from the calculus certain statements which are compared
with experience. These experimental comparisons form points
of contact between the calculus and reality, and from these
points *empirical* meaning flows upward – inundating the
theoretical terms of the system. To use another analogy of

Braithwaite's we could say that the zip-fastener is joined together at the low level where comparison with experience takes place and is then zipped up to the high-level theoretical terms.

There is no problem about giving the term 'probability' logical meaning because we can set up a mathematical calculus of probability in the usual way without difficulty. The problem is how to give the term 'probability' empirical meaning by attaching the mathematical calculus to reality at the experimental level. Braithwaite proposes that we adopt a k-rule of rejection which tells us that a probability statement is to be rejected – at least provisionally – under certain circumstances. This gives us an implicit definition of the term 'probability' of the sort we require.

The formulation of the k-rule of rejection proceeds as follows. Suppose we wish to give empirical meaning to the statement s that the probability of event A relative to conditions $\mathfrak{S}_s$ is p. (We are here translating Braithwaite's ideas into our own terminology.) Suppose $\mathfrak{S}_s$ is repeated n times and A occurs m times. We then deduce from the mathematical calculus (assuming implicitly that the repetitions are independent) that

$$\text{prob}\,(|m/n - p| > \epsilon) < pq/n\epsilon^2$$

where ϵ is an arbitrary constant > 0, and $q = 1 - p$. This is in fact just a special case of the Chebichev inequality – a standard theorem of the mathematical calculus. Now set $k = pq/n\epsilon^2$. We then have that with a probability of more than $1 - k$ that m/n will lie in the interval $[p - (pq/nk)^{1/2}, p + (pq/nk)^{1/2}]$. Our k-rule of rejection accordingly states that we should provisionally reject the statement s if m/n in fact lies outside this interval. This rule of rejection gives the required empirical meaning to s.

The value of k in the k-rule is of course arbitrary to a considerable degree. On Braithwaite's view we give a different empirical meaning to probability statements depending on the value of k we choose. Thus the intuitive notion of probability is explicated by a continuum of k-probabilities. This is a curious feature of Braithwaite's theory, and the theory is criticized by Hacking on this ground (cf. 1965, pp. 114–17). Perhaps Braithwaite could avoid this difficulty by fixing on a particular k-value – say 5 per cent.

Another point is that Braithwaite's use of the Chebichev inequality does not seem very satisfactory. We have

$$\text{prob}\,(m/n \notin [p - (pq/nk)^{1/2},\, p + (pq/nk)^{1/2}]) < k,$$

but this probability may in some cases be very much less than k so that the value k is not very significant. In fact we have that for large n, m/n is approximately normally distributed and therefore

$$\text{prob}\,(m/n \notin [p - \lambda_k(pq/n)^{1/2},\, p + \lambda_k(pq/n)^{1/2}]) \doteqdot k \qquad (*)$$

where λ_k is the k per cent point of the normal distribution. Accordingly it seems better to use a rejection rule based on this interval and we will adopt this version of the theory in future. In fact Braithwaite himself makes this point (1953, p. 168), but does not include it in the main formulation of his theory.

We now come to a more significant feature. The statement (*) can be interpreted as saying that there is a probability of approx. k per cent of wrong rejection when we use a k-rule. Thus the rule, which is designed to explain the meaning of probability, contains a clause about probability, and would therefore seem to give a circular explanation. As Braithwaite puts it (1953, p. 155): 'So, as it may well be said, it is circular to explain the original probability statement in terms of the circumstances under which it would be rejected, when the rejection itself has to be explained in terms of probability.'

Braithwaite's solution is as follows. Consider the original k-rule. It is explained in terms of a probability statement s' to the effect that there is a k per cent probability of wrong rejection. We now give s' meaning by a further k-rule. This k-rule involves a new probability statement s'' which is given meaning by yet a third k-rule etc. Thus the original probability statement is ultimately given its empirical meaning by an infinite hierarchy of k-rules.

(ii) Braithwaite's theory of probability: criticism
Such then are the main features of Braithwaite's theory. Let us now compare it with our own view. The first difference is that we introduce our falsifying rule to solve a *methodological* problem whereas Braithwaite introduces his to solve a problem about *meaning*. Braithwaite is seeking to give an implicit definition

of probability statements in terms of his infinite hierarchy of *k*-rules. In contrast our own problem arose like this. We accepted Popper's theory that scientific statements must be falsifiable, and then were faced with the difficulty that probability statements were both scientific and impervious to falsification. The solution was that probability statements, although strictly unfalsifiable, could be used as falsifiable statements. Our problem was then to formulate explicit instructions as to how probability statements were to be used in a falsifiable manner. These instructions were the falsifying rule (F.R.P.S.). So the whole problem and the solution to it did not bring in questions of meaning. The problem was rather one about how we should handle probability hypotheses, i.e. a methodological problem. The second difference concerns the constants involved in the F.R.P.S. In the formulation of our own F.R.P.S. we had a constant *k* similar to Braithwaite's. (Indeed I adopted the letter '*k*' to signify my indebtedness to Braithwaite on this point.) However, as well as *k*, we found it necessary to introduce *l* (the relative likelihood of the rejection class) and there is no such factor in Braithwaite's theory. Thirdly and lastly, Braithwaite gets involved in an infinite regress of *k*-rules whereas there is no such regress in our own theory. I think it will be agreed that an infinite regress should be avoided if it can be. But can it validly be avoided? Should not we too have an infinite regress? This is evidently a question which it is most important for us to tackle. We will now examine these three differences in rather more detail.

The contrast between Braithwaite's *meaning* and our *methodological* approach is in fact less significant than it might at first seem. Consider in this context Wittgenstein's thesis (1953, p. 20, §43): 'For a *large* class of cases – though not for all – in which we employ the word "meaning" it can be defined thus: the meaning of a word is its use in the language.' Whether Wittgenstein is right that the meaning of a word can be *identified* with its use, we will not here discuss. It seems however almost indubitable that the use of a word at least partly determines its meaning. Thus if we set up a methodological rule stating how a particular calculus is to be used in practice, we are thereby partly determining the meaning of the terms in the calculus. If we used the calculus according to a different methodological rule, we would have given it a different meaning. In particular

13

our F.R.P.S. can be considered as in part determining the meaning of probability statements.

But now we come to an important difference with Braithwaite. Braithwaite's *k*-rule of rejection does not in fact give a set of general instructions about how probability statements are to be used. It does not like ours underlie any arbitrary statistical test. Instead of being a common feature of every statistical test, it is in effect a *particular* statistical test based on the Chebichev inequality (or in our version on the normal approximation). Moreover Braithwaite does not think that this particular test should always or indeed ever be used in practice. He says (1953, p. 161, footnote):

> It is important to bear in mind throughout in reading this chapter that the series of rejection tests is proposed for use in an explanation of the *meaning* of a probability statement, and is not proposed as the best method for deciding whether or not a suggested value for a probability parameter should be rejected, still less as the best method for discriminating between two or more suggested values.

In the next chapter Braithwaite gives a general account of how he thinks statistical hypotheses *ought* to be dealt with in practice, but this account is in terms of decision theory and not in terms of a falsifying rule. However, if statistical hypotheses are dealt with in practice in accordance with decision theory rather than in accordance with an F.R.P.S., it is the rules of decision theory which give 'practical' or 'empirical' meaning to probability. It seems to me that Braithwaite has to make a choice. *Either* he must reformulate his rejection rule to make it a general instruction for handling statistical hypotheses – in this case it can genuinely be said to give meaning to probability – *or* he should abandon his rejection rule and say that the instructions of decision theory give meaning to probability. What does not seem satisfactory is the claim that a specific and hypothetical test, which on his own confession may never be used in practice, can give empirical meaning to probability.

The fact that Braithwaite's *k*-rule is a 'specific test' rather than a 'general instruction underlying all statistical tests' explains why he does not need to introduce the *l*-factor. In effect Braithwaite, implicitly but not explicitly, chooses the falsification class in his particular case to have low *l*-value (relative likelihood). Since he is not formulating a rule which can apply in all cases, he can single out the falsification class by a particular specification rather than by a series of general requirements. Thus he does not consider whether we might have a falsification class of the form $(p - \delta, p + \delta)$, but stipulates that it be of the form $[0, p - \lambda_k(pq/n)^{1/2}) \cup (p + \lambda_k(pq/n)^{1/2}, 1]$. Such a stipulation will not be adequate in general since we may be dealing in our test with a non-normal falsifiable distribution. To cover all such cases it is necessary, as we have shown, to have a *k*-, *l*-rule of rejection and not just a *k*-rule of rejection.

We now come to the question of whether an infinite regress of *k*-rules is really necessary. I claim that it is not. The argument which led to the infinite regress ran as follows. The explanation of the *k*-rule of rejection involves a statement to the effect that there is a probability of *k* per cent that the hypothesis will be rejected when it is in fact true. This probability statement must in its turn be explained by a *k*-rule and so we are involved in a regress. Now consider by way of comparison the use of a set of axioms to give an implicit definition of one or more mathematical concepts. Suppose, for example, we are defining 'point' and 'line' by Hilbert's axioms. Now Hilbert's axioms do of course make statements about points and lines. Could we not argue as follows? 'These axioms are supposed to be defining "point" and "line" and yet they mention points and lines. If we are not to be circular we must explain the meaning of "point" and "line" at such occurrences by more axioms, and thus we are led to an infinite regress.' Such a line of reasoning is plainly absurd and rests on a misunderstanding of the process of implicit definition.

Now take the case of probability. This time our implicit definition consists of a set of axioms + an F.R.P.S. But just as we do not have to explain the meaning of probability when it occurs in the axioms, so we do not have to explain its meaning when it occurs in the formulation of the F.R.P.S. In both cases the occurrences are part of the implicit definition. In this way we can, fortunately, avoid Braithwaite's infinite regress.

Some uneasiness may still be felt because (i) we set out to give meaning to statements of the form: 'There is a *p* per cent probability of such and such an event' and (ii) our explanations apparently contain a statement of exactly this form with

$p = k/100$. However it is worth noting that we can formulate a k-rule of rejection as follows: 'Reject statement s provisionally if observed result lies in interval I_k.' This formulation does not involve any statement to the effect that there is a k per cent chance of wrong rejection. If now the set of axioms + the k-rule of rejection really give an adequate implicit definition of probability, then all statements of the form: 'There is a p per cent probability of such and such an event' acquire meaning, *including* the statement that there is a k per cent probability of wrong rejection.

The argument which leads to the infinite regress is strictly speaking wrong, but it is closely connected with a genuine difficulty. To see what this is, observe that the statement: 'There is a k per cent probability of wrong rejection' is not involved in the formulation of the k-rule of rejection, but would be involved in an attempt to justify that rule. The problem in effect lies not in formulating the falsifying rule but in justifying it. We could no doubt invent all sorts of strange falsifying rules. If we conjoined one of these to the probability axioms, we would perhaps obtain a system which implicitly defined *some* concept of probability. No doubt this concept would differ from our usual one, and probably it would be of little interest. However, the possibility of such alternatives leads us to ask what reasons can be given to justify the choice of our own particular falsifying rule. This is of course the central question of the present chapter to which we have been naturally led by a consideration of Braithwaite's theory. Braithwaite's infinite hierarchy of rules is no help towards solving this problem. It merely gives an unnecessarily complicated formulation of a falsifying rule without in any way justifying such a rule. We could always ask: 'Why should we use Braithwaite's hierarchy to guide our statistical procedures?'

(iii) Possible ways of justifying the F.R.P.S.

How then can we try to justify the adoption of a falsifying rule? There are, so far as I can see, three possible ways: (1) by its agreement with intuition, (2) by the practical success obtained when using it and (3) by a proof that it can 'consistently' be adopted. We will now consider these three approaches in turn.

We have already given our general philosophy of arguments

which involve an appeal to intuition. Our 'intuitions' regarding a particular notion may be supposed to arise in the following manner. When a 'new' concept such as probability or mass is introduced, it will nearly always be found that cruder and more qualitative versions preceded the more precise idea. Moreover certain theories and assumptions would naturally be held regarding these cruder versions. Now some assumption regarding the new, precise version of the concept is 'in agreement with intuition' if it is justified by an old theory or assumption involving the cruder version. Thus arguments from intuition are justified by the 'general principle of correspondence' which states that our new theories should always agree in large measure with our old. At the same time we see that arguments from intuition are never very powerful because new theories do also have to differ from the old for the sake of greater elegance, or better agreement with the facts.

A typical argument from intuition concerned the assumption 'mass (Sun) $\gg$ mass (Earth)' needed to test Newton's theory. Crude pre-Newtonian notions of 'mass' or 'matter' would have led to the truth of this proposition, and thus it could validly have been said to agree with intuition. Similarly a pre-formal notion of probability suggests that it is reasonable to neglect a small probability in comparison with a large one. The other arguments used in Chapter 9 when we were formulating our F.R.P.S. could equally be justified 'by intuition'. In short, I conclude that our F.R.P.S. does agree with intuition and that this gives some kind of justification for adopting it. However this justification is far from completely convincing. After all, the F.R.P.S. which perhaps most naturally suggests itself to intuition is the naive R.1: 'Regard H as falsified, if the observed result has low probability given H.' Moreover this rule proved unacceptable, and the same might be true of our final F.R.P.S. despite its intuitive plausibility.

(iv) The practical success but inconsistency of the F.R.P.S.

We are thus led to our second line of approach in terms of the practical success of the rule. This, I believe, yields the strongest and most convincing arguments in favour of the rule. After all, how do we justify the adoption of a set of axioms for an empirical theory? The method is to derive a set of observable results from these axioms and check that these results agree with experience. In the case of probability theory we have to adopt a set of axioms *and* a falsifying rule; but otherwise the case is the same. From this system we can, as has been shown, derive more precise versions of the two empirical laws of probability theory and these more precise versions are found to agree with experience. Suppose we retained the probability axioms but adopted a different falsifying rule – R' say. In these circumstances either we would not be able to derive the two empirical laws or we would be able to derive some further results which did not agree with experience. In either case we would have to admit that R' had not proved its mettle in practice.

So then our main argument for adopting the F.R.P.S. is the empirical success of the system consisting of it and the probability axioms. This argument should, however, be buttressed by a consideration of whether we can consistently adopt the F.R.P.S. or whether its use in practice leads to contradiction. This brings us to our third line of argument. Suppose we do adopt the F.R.P.S. Then we have that $\text{prob}(C) = k < k_0$ where C is the falsification class. The value k will vary in different applications. Let us suppose, however, that k is constant and nearly equal to k_0. This will in fact be approximately true in practice. Let us now consider a large number n of tests where the hypothesis H is in fact true. Suppose H is falsified (wrongly) in m of these tests. Since in each instance we have: 'prob (wrong rejection) = a constant k', the law of stability of statistical frequencies (which it must be remembered was derived using the F.R.P.S.) gives us that

$$m/n \doteqdot k/100$$

where the accuracy of the approximation is given by the more precise formula of Part II.

Let us consider what this result means. We have assumed (i) the axioms of probability and (ii) the F.R.P.S. with the related notions of f-deducibility etc. From this system S we infer that the F.R.P.S. will lead to a wrong falsification of a true statistical hypothesis in approximately k per cent of the cases where it is applied to such hypotheses. In other words, if we are right to rely on the F.R.P.S., we are right to believe that it will give us the wrong answer in k per cent of cases of a certain type.

In other words we *cannot* consistently adopt the F.R.P.S. The rule will inevitably lead to inconsistency.

It is most important to recognize this inconsistency because it shows that we must always be on our guard when handling probability statements in accordance with the F.R.P.S. and that we can easily be led astray. We have already noted one example of this. If we study the observed values of a set of random variables, we can always design a test which will falsify the underlying statistical hypothesis – be it ever so true. We have seen that certain commonsense considerations can enable us to avoid gross errors of this sort, but, as we also saw, this is no merely theoretical difficulty but one which practising statisticians must take care to avoid. The inconsistency in the F.R.P.S. should therefore warn us to take care. Is it, however, fatal to the whole approach?

I would argue that it is not. The first point to note is that even inconsistent deductive systems in pure mathematics are by no means unworkable. Consider Frege's set theory for example. Russell's paradox showed it to be inconsistent, and yet most of the results derived by Frege were, relative to modern standards, correct and had been given correct proofs (or at least proofs which needed only a little correcting). It would be perfectly possible even now to use Frege's set theory provided care was taken to avoid the kind of reasoning which led to the paradoxes. Indeed, many mathematicians, when arguing informally, do more or less this. Of course corrected set theories (e.g. Zermelo–Fraenkel set theory) have been invented, but even they might turn out to be inconsistent. If they were inconsistent, however, there is little doubt that much of the reasoning which had been conducted in them would remain valid with at most a little correction.

Of course in the case of probability theory and the F.R.P.S. we are not concerned with a purely logical inconsistency but with a kind of practical inconsistency. However the analogy with inconsistencies in set theory shows that the inconsistency need not prove fatal provided we proceed with care. An important difference is that the inconsistency in the F.R.P.S. cannot, as far as I can see, be removed; but because it is an inconsistency which arises only in applications of the theory it is correspondingly less serious. After all, whenever we apply mathematical theories to real situations, there are always numerous possible sources of error, and the inconsistency in the F.R.P.S. only adds one more to this number.

To see this let us suppose we are testing a true deterministic hypothesis *H*. We might erroneously regard it as falsified for a variety of reasons. The result deduced from *H* and compared with experience could be wrong simply through mathematical error or because quantities had been neglected in the deduction which in fact had a significant effect. Certain motions of the Moon were wrongly regarded as contradicting Newton's theory early in the eighteenth century for exactly this last reason. Again, the experimental result obtained could be erroneous either through a simple mistake in observation or through a mistaken theoretical interpretation of the observations. Numerous experiments have had to be reinterpreted because some 'factor' (the expansion of the glass, the refraction in the atmosphere etc.) was forgotten. All these sources of error can and indeed have led to the erroneous falsification of true hypotheses in the deterministic case. In the statistical case we merely add an extra possibility of error which arises because of the use of the F.R.P.S.

Similar considerations apply if we look at it from the other point of view. Suppose we are explaining or predicting some result *e* from a statistical hypothesis *H*. Proceeding in accordance with the F.R.P.S. we 'f-deduce' a proposition of the form '$\xi \in A$' which is equivalent to *e*. Now of course even if *H* is true, there is a *k* per cent chance that $\xi \notin A$ and our prediction will be erroneous. Once again we seem to have a remarkable possibility of error where nothing corresponding arises in the deterministic case. Once again, however, the sharp difference is misleading because there are many possible sources of error in the deterministic case as well. Leaving aside the question of faulty mathematical deductions, there is still the ever-present possibility that we are using a false hypothesis to make the prediction. It may, for example, be a hypothesis which has worked in other cases but breaks down in the particular case under consideration. So, as before, we are not adding a source of error to an otherwise perfect method, but one source of error to a method which is already liable to many.

We can now sum up our general conclusions regarding the

justification of the F.R.P.S. First of all, it can certainly be said to be intuitively plausible, but this is a weak argument in its favour. The only really strong argument for it is that by adopting it together with the axioms of probability we obtain a system which works in practice, in the sense that we can derive from it results which agree with experience. Ideally we would like to supplement this by a proof that we can consistently adopt the F.R.P.S. However, it turns out that we are in fact inconsistent in adopting it, in the sense that if we rely on the F.R.P.S. it can be shown that it will give us the wrong result in k per cent of cases of a certain type. This inconsistency should put us on our guard against the ever-present possibility of error when we use the F.R.P.S. – a possibility which practising statisticians have indeed recognized. However the inconsistency does not seem to be fatal to the whole approach. It merely adds one more source of error where many already exist. To these general remarks I will now add two rather more particular points.

(v) The F.R.P.S. in relation to the theory of errors
The first of these considerations is yet another attempt to minimize the differences between the deterministic and the statistical cases. We proceed by relating the problem to that of 'intervals of imprecision', a course followed by Popper (1934, § 68). Suppose we are testing a deterministic hypothesis H. We might in some particularly simple cases deduce that, given H, a certain measurable quantity x should have a value x_0. We would then measure x to see whether it did indeed equal x_0 or not. More usually we might deduce, given H, that two measurable quantities λ and μ were linearly related $\lambda \propto \mu$. We would then measure a number of pairs of values of λ and μ, $(\lambda_1, \mu_1) \dots (\lambda_n, \mu_n)$ say, and see whether these pairs did indeed fall on a straight line. Now would we regard H as falsified if x differed by any quantity, however small, from x_0, or if the curve joining the (λ_i, μ_i) departed in any degree whatsoever from linearity? Of course not. Indeed we would expect x to differ from x_0 and would be surprised if the two quantities were experimentally indistinguishable, or if the (λ_i, μ_i) lay *exactly* on some line. The difference from x_0 or the departure from linearity would be attributed to 'experimental error'. But now could we say that the experiment agreed with H however much x

differed from x_0 or however randomly the points (λ_i, μ_i) were scattered on the plane? Could we not merely say that the experimental errors had been large in such cases? Once again we would of course reject such an absurd view. In fact we would regard the hypothesis as confirmed if we obtained a result sufficiently near x_0 and falsified if the result was too far away. In other words we surround x_0 by a certain, approximately defined, 'interval of imprecision' $[x_0 - \phi, x_0 + \phi]$ and regard H as confirmed if the result is in the interval and falsified otherwise. Similarly, in the (λ, μ) case we take a certain band in the plane which we regard as a sufficient approximation to a straight line. Now this procedure is surely very similar to the adoption of a falsifying rule. In both cases we could in theory allow any divergence however large. Yet in practice we draw the line in an admittedly somewhat arbitrary fashion, and only permit divergences up to a certain point. In both cases this decision renders the underlying hypothesis falsifiable.

The analogy is heightened if we consider that experimental errors can be treated statistically. Let us show this in the simple case where we predict $x = x_0$. The (λ, μ) example is similar but involves considerations of regression. Various measurements of the quantity concerned can be considered as independent trials. These trials can be represented by $x - x_0$ now considered as a random variable, whose range can be taken as the whole real line. Intuitively the value of $x - x_0$ on a particular trial shows the degree of error in the measurement, i.e. the degree to which it deviates from the predicted value of x_0. If we now assume that this error is the sum of a large number of mutually independent elementary errors, we obtain that $x - x_0$ is distributed approximately normally about the value 0. Of course the assumption behind this deduction is not always very plausible and indeed other distributions of the error random variable often agree better with observation.

The actual form of the distribution is not important for our purposes however. Our question is this. Suppose we assign some distribution D to the error random variable $x - x_0$. How does this more sophisticated statistical treatment tie in with the usual procedure of assigning an interval of experimental imprecision $[x_0 - \phi, x_0 + \phi]$ to x_0? The answer is that D must be a falsifiable distribution whose acceptance region A is the interval of imprecision. But now we see that the selection of an interval of imprecision and the application of the F.R.P.S. become in this case exactly equivalent. Thus the adoption of the F.R.P.S. can be seen as a natural generalization of the way in which errors are treated when testing ordinary deterministic theories. Indeed we could formulate the difference between statistical and deterministic science in the following fashion. In deterministic science no statistical considerations appear in the laws and theories themselves and only come in when these laws and theories are tested. In statistical science probability enters the laws as well.

(vi) Resolution of a problem connected with the F.R.P.S.

My last 'general consideration' regarding the F.R.P.S. is an attempt to remove a paradox which apparently results from an adoption of that rule. Consider again the standard coin-tossing example where we suppose the tosses are independent and prob (heads) = prob (tails) = $\frac{1}{2}$. Suppose we toss the coin 1,000 times and get tails each time. Now consider two random variables ξ and η. ξ is the standard relative frequency of heads. It has of course a falsifiable distribution and the result in question which may be written $\xi = 0$ falsifies the hypothesis. We now define η as follows. Consider the set of $2^{1,000}$ permutations of heads and tails. Arrange this set in some order beginning with $\langle 0, 0, \ldots, 0 \rangle$ and ending with $\langle 1, 1, \ldots, 1 \rangle$ where 0 stands for tails and 1 for heads. Number the set 1, 2, ..., $2^{1,000}$. Let us say we obtain the result $\eta = n$ if the nth of these permutations occurs. Then η has the uniform and hence unfalsifiable distribution

$$\text{prob} \, (\eta = n) = 1/2^{1,000} \qquad (1 \leqslant n \leqslant 2^{1,000}).$$

Consequently whatever value of η we obtain it is compatible with the hypothesis. In particular, if we obtain $\eta = 1$ (i.e. all tails) the result is compatible with the hypothesis. In other words our F.R.P.S. has the following consequence. Suppose we obtain the result 1,000 tails. If we interpret this as the value 0 for the r.v. ξ our hypothesis is falsified. If we interpret it as the value 1 for the r.v. η the result is compatible with the hypothesis. Yet the observations are the same in the two cases. How can a mere mathematical manipulation change them from falsifying the hypothesis to being compatible with it?

This result seemed to me distinctly puzzling – if not para-doxical – for a long time, until what I believe is the correct solution was given to me by my friend J. Worrall. I will now expound Worrall's idea. His main point was that theories are falsified by facts, and 'facts' are not 'crude observations' but 'observations interpreted in the light of theories'. Thus a thermodynamic theory is not refuted by observing that 'the tip of a greyish column embedded in glass is above a black mark next to which the figure 100 is inscribed'. It might however be refuted by the fact that the liquid X is boiling at more than 100°C. Yet this latter 'fact' is merely an interpretation of the previous 'observations'. So in the deterministic case certain 'observations' may only refute a theory if interpreted in a certain way. Similarly, we should not after all be surprised that a certain result refutes a statistical hypothesis if interpreted as '$\xi = 0$' but not if interpreted as '$\eta = 1$'.

CHAPTER ELEVEN

The Neyman-Pearson Theory

An examination of the Neyman–Pearson theory is not for us an option but on the contrary a most pressing necessity. The reason is this. As we shall show in a moment, the Neyman–Pearson theory in certain circumstances gives results which directly contradict our own approach based on the falsifying rule. As the Neyman–Pearson theory is still the generally accepted theory of testing statistical hypotheses, this fact could be used as a strong argument against our views. Our reply to this argument will be to criticize the Neyman–Pearson theory and to suggest that it, rather than our falsifying rule, should be given up. First, however, it would seem desirable to summarize the main points of the Neyman–Pearson theory as briefly as possible.

(i) Outline of the Neyman–Pearson theory
I will begin with an example. Suppose we want to test the hypothesis H_0 that a certain random variable ξ is normally distributed with mean μ_0 and standard deviation σ_0. We suppose our evidence consists of n independent observed values of ξ, $x_1, \ldots, x_n$ say (the sample). $(x_1, \ldots, x_n)$ is considered as a single point $\mathbf{e}$ or $\mathbf{x}$ in an n-D space of possible values – the evidence space (or sample space) E. The testing problem as conceived by Neyman and Pearson is that of choosing a subset C of E s.t. if the observed value $\mathbf{e} \in C$ then we should reject H_0. C is known as a critical region. Neyman and Pearson suggest that C should be chosen so as to minimize two types of error. (These two types are distinguished in their very first joint paper, cf. Neyman and Pearson, 1928, p. 3.) Type 1 error is the error of rejecting H_0 when it is in fact true. Type 2 error is the error of accepting H_0 when it is in fact false.

The treatment of type 1 error is fairly easy. H_0 induces a probability measure over E. Thus if S is any (Borel) subset of E, we can calculate $\mathrm{prob}(S, H_0)$. In particular we can find $\mathrm{prob}(C, H_0)$. We control type 1 error by choosing a low value for this. If $\text{prob}(C, H_0) = \epsilon$, we say we are using a test of *size* ϵ, or of *level of significance* ϵ. In general it is assumed that $\epsilon = 5$ per cent, i.e. we use a 5 per cent level of significance. The trouble here is that we can choose a large number of sets C s.t. $\text{prob}(C, H_0) = \epsilon$. Of these many will be intuitively unsatisfactory as critical regions, and to eliminate these unsatisfactory ones and obtain a good critical region, we must, according to Neyman and Pearson, consider type 2 error.

It is in the introduction and treatment of type 2 error that the novelty of the Neyman–Pearson theory consists. Neyman–Pearson assume that we can introduce a set of alternative hypotheses such that if H_0 is false, one of these alternatives is true. In general we obtain these alternatives by varying the parameters of the original hypothesis. Thus in our example we might take as alternatives $H_{\mu\sigma} = $ 'ξ is normally distributed with mean μ and standard deviation σ, where μ is arbitrary and σ is arbitrary but > 0'. We might instead take alternatives $H_{\mu\sigma_0}$ for arbitrary μ but fixed σ_0, or even some subset of this. We can now consider the probability of an arbitrary (Borel) subset S of E when an alternative, H_α say, is true. Denote this by $\text{prob}(S, H_\alpha)$. To reduce type 2 error we have to make $\text{prob}(C, H_\alpha)$ as large as possible for alternatives H_α, because $\text{prob}(C, H_\alpha)$ represents the probability of rejection when the alternative H_α is in fact true. We will now give a general formulation, and explain how this last requirement can be made more precise.

We will begin then with a set Ω of hypotheses. Each hypothesis specifies completely the distribution of a certain random variable ξ. It will be supposed that each hypothesis is specified by the values of a set k of parameters $\mu = (\mu_0, \ldots, \mu_k)$. In the example above these were the mean μ and standard deviation σ. A point $\omega \in \Omega$ will be called a *simple* hypothesis, a set of points in Ω a *composite* hypothesis. We will first consider the problem of testing a simple hypothesis against the alternatives in Ω. At the end we will examine the modification necessary if we are testing a composite hypothesis. As before we will assume an evidence space (or sample space) E, and take the problem to be that of choosing a $C \subset E$ (a critical region) for testing our hypothesis, μ_0 say. We often call a test 'the test C', meaning by that 'the test whose critical region is C'. It is required that C should have some fixed *size*, ϵ say, i.e. $\text{prob}(C, \mu_0) = \epsilon$. We further define prob (C, μ) for variable μ as the power function of the test. We reduce the type 2 error by making prob (C, μ) as large as possible for $\mu \neq \mu_0$. Thus, speaking loosely, the Neyman–Pearson theory requires maximum power for a given size.

The first important result of the theory is the fundamental lemma of Neyman and Pearson. Suppose we have only one alternative hypothesis, μ_1 say, and that both μ_0 and μ_1 induce continuous probability distributions in E with probability density functions $f(\mathbf{x}, \mu_0)$ and $f(\mathbf{x}, \mu_1)$. Suppose we take all our tests to be of size ϵ. That is to say we require prob $(C, \mu_0) = \epsilon$, where C is a putative critical region. Then we can find a test K of maximum power. That is to say if C is any other test of size ϵ, then

$$\text{prob} \, (K, \mu_1) \geqslant \text{prob} \, (C, \mu_1).$$

In fact we can define K as the set of $\mathbf{x}$ s.t.

$$f(\mathbf{x}, \mu_1) \geqslant c f(\mathbf{x}, \mu_0)$$

for some $c > 0$ and s.t. prob $(K, \mu_0) = \epsilon$. We then have:

Fundamental Lemma of Neyman and Pearson
If K is defined as above, then prob $(K, \mu_1) \geqslant \text{prob} \, (C, \mu_1)$ for any C s.t. prob $(C, \mu_0) = \epsilon$, i.e. K is the test of maximum power for given size ϵ.

This gives a solution to the problem for a single alternative. If we have a number of alternatives, the situation is not so simple. It may happen however that we can find a set $K \subset E$ of size ϵ s.t. if C is any other set of size ϵ, then

$$\text{prob} \, (K, \mu) \geqslant \text{prob} \, (C, \mu)$$

for *all* alternatives μ. Such a set is called a uniformly most powerful test or UMP test. If it exists it is considered optimal in the Neyman–Pearson theory. If such a test does not exist (as is in general the case) we can sometimes find tests satisfying weaker conditions. Let us call a test C of size ϵ *unbiased* if there is no $\mu_1 \neq \mu_0$ s.t. $P(C, \mu_1) < \epsilon$. We then say a test is uniformly most powerful unbiased (or UMPU) if it is uniformly most powerful *among unbiased tests of a given size* ϵ. More formally a test K is UMPU of size ϵ, if, for any other C which is an unbiased test of size ϵ, we have

$$\text{prob} \, (K, \mu) \geqslant \text{prob} \, (C, \mu).$$

Sometimes we can find a UMPU test where no UMP test exists.

We will now say a word about how this should be modified if we are testing not a simple hypothesis H_0, but a composite one (H_M say) which only specifies that $\mu \in M$ where M is some subset of Ω. This can occur e.g. when we want to test whether ξ is normally distributed with zero mean, but do not want to specify the standard deviation σ which can thus take any value $0 < \sigma < \infty$. The trouble here is that we cannot define the size of an arbitrary test (C say) because prob (C, μ) may be different for different $\mu \in M$. Suppose however prob $(C, \mu) = \epsilon$ for all $\mu \in M$. C is then said to be *similar to the sample space*. If we confine ourselves to tests which are similar to the sample space, we can then define UMP tests and UMPU tests just as before.

We will conclude our summary of the Neyman–Pearson theory by giving some simple examples. Suppose η is normally distributed with zero mean, and standard deviation 1. Let η_p be s.t. prob $(|\eta| \geqslant \eta_p) = p$. As before we will call η_p the p per cent point of the normal distribution, and similarly with other distributions. Now suppose we are testing the hypothesis $H_{\mu_0 \sigma_0}$ that a certain random variable ξ is normally distributed with mean μ_0 and standard deviation σ_0. We will suppose σ_0 to be 'known' so that only alternatives which assign normal distributions to ξ with different means μ but the *same* standard deviation ($H_{\mu\sigma_0}$ say) will be considered. Let $\bar{x} = (x_1 + \cdots + x_n)/n$. Then of course $\eta = (n)^{1/2}(\bar{x} - \mu_0)/\sigma_0$ is normally distributed with zero mean and unit standard deviation. It can be shown that the test which consists in rejecting $H_{\mu_0 \sigma_0}$ when

$$\bar{x} \geqslant \mu_0 + \eta_{2\epsilon}\sigma_0/(n)^{1/2}$$

is an UMP test of size ϵ against the set of alternative hypotheses $H_{\mu\sigma_0}$ with $\mu > \mu_0$. Similarly the test which consists in rejecting $H_{\mu_0 \sigma_0}$ when $\bar{x} \leqslant \mu_0 - \eta_{2\epsilon}\sigma_0/(n)^{1/2}$ is an UMP test of size ϵ against the set of alternatives $H_{\mu\sigma_0}$ with $\mu < \mu_0$. If we consider the full set of alternatives $H_{\mu\sigma_0}$ with $\mu \neq \mu_0$ there is no UMP test. But the test which consists in rejecting $H_{\mu_0 \sigma_0}$ if $|\bar{x} - \mu_0| \geqslant \eta_\epsilon \sigma_0/(n)^{1/2}$ is an UMPU test of size ϵ against this set of alternatives.

This example shows where the Neyman–Pearson theory differs from our own approach. For us the last or '2-tailed' test is quite valid, but the 1-tailed tests which consist in rejecting $H_{\mu_0 \sigma_0}$ when $\bar{x} \geqslant \mu_0 + \eta_{2\epsilon}\sigma_0/(n)^{1/2}$ or $\bar{x} \leqslant \mu_0 - \eta_{2\epsilon}\sigma_0/(n)^{1/2}$ are invalid. To see this observe that with such critical regions C, we can certainly find points $\bar{x}$ in the acceptance region s.t. $l(\bar{x}) < l(C)$. (Points in the left-hand (resp. right-hand) tail of the normal curve can be chosen for this.) This contradicts even the weaker preliminary version R.3 of our falsifying rule and thus evidently the final stronger version. On the other hand such 1-tailed tests are UMP in a certain class of alternatives, and thus would presumably be valid under certain circumstances on the Neyman–Pearson theory. This justifies our claim that the two theories lead to different results under certain circumstances, and thus heightens the interest in comparing them.

This first example is a trifle artificial. Let us therefore consider testing the composite hypothesis $H_{\mu_0 \sigma}$ for any $\sigma > 0$. (We use the notation of the previous example.) If we set

$$s = 1/n \sum_{i=1}^{n} (x_i - \bar{x})^2$$

we have that $t = (n-1)^{1/2} (\bar{x} - \mu_0)/s$ has the t-distribution with $n-1$ degrees of freedom. Then as before the test which consists in rejecting $H_{\mu_0 \sigma}$ if $t \geqslant t_{2\varepsilon}$ is UMP of size ε against the alternatives $\mu > \mu_0$, σ arbitrary > 0; similarly $t \leqslant t_{2\varepsilon}$ is UMP if size ε for $\mu < \mu_0$, σ arbitrary > 0. If we consider the full set of alternatives $H_{\mu\sigma}$ μ arbitrary σ arbitrary > 0, there is no UMP test. But the test which consists in rejecting $H_{\mu_0 \sigma}$ if $|t| > t_\varepsilon$ is UMPU of size ε against this set of alternatives. Here of course t_ε is the ε per cent point of the t-distribution with $n-1$ degrees of freedom.

(ii) The Neyman–Pearson theory without an F.R.P.S.

Our first criticism of the Neyman–Pearson theory proceeds by attempting to relate this theory to the notion of a falsifying rule for probability statements. There seem to be two possibilities. Firstly, it could be held that the Neyman–Pearson theory renders a falsifying rule superfluous; that we should have the Neyman–Pearson theory but *without* an F.R.P.S. Secondly, the directly opposed view could be held that the Neyman–Pearson theory actually embodies a falsifying rule, or that we can easily formulate a falsifying rule in terms of the Neyman–Pearson theory. I will consider these two positions in turn.

The concept of the Neyman–Pearson theory *without* a falsifying rule is proved incoherent by the following argument. The Neyman–Pearson theory allows us to test out a given

14

hypothesis against a set of alternatives, and to reject or accept the hypothesis. Suppose now that we test out a given hypothesis H_0 by the Neyman–Pearson procedures, and we come to the conclusion that H_0 should be accepted. My argument is that without a falsifying rule we cannot then do anything with H_0, the hypothesis we have accepted. Thus the whole testing procedure becomes rather pointless. This argument can in fact be used to attack quite a number of statistical theories. Too much attention is focused on the problem of selecting one statistical hypothesis out of a number of alternatives, and the question of what uses the hypothesis will be put to once selected is forgotten. Often, as in the present case of the Neyman–Pearson theory without an F.R.P.S., it actually becomes impossible to make any use of a hypothesis which has been 'accepted'.

In order to justify this last statement I will begin by dividing the possible uses of a hypothesis into two classes, theoretical and practical, and will consider these two cases separately. First, then, let us consider the possible theoretical uses of a hypothesis. From a theoretical point of view there are, as far as I can see, only two uses of a hypothesis: to explain known results, and to predict new results. For simplicity I will confine myself to the explanation and prediction of observable phenomena (i.e. basic statements) and will leave aside possible explanations of higher-level laws. If we have a falsifying rule it is of course very easy to explain or predict some basic statement e. In either case we f-deduce e from the hypothesis H. But how can we use a statistical hypothesis H to make a prediction or to give an explanation without a falsifying rule? As far as I can see it would be impossible to do so. In order to get an explanation we need to form a connection between a probability hypothesis H and a basic statement e, presumably a frequency statement. In other words we need to establish a connection between probability and frequency, and as my earlier discussion of Part II indicated, some kind of falsifying rule is necessary for this. Thus if we have the Neyman–Pearson theory without a falsifying rule, we can use it to test out and render acceptable a hypothesis H_0. But we cannot unfortunately then use H_0 to explain any observed phenomenon or to predict any observable phenomenon. From a theoretical point of view H_0 is useless.

Against this it might be objected that the important question is not the theoretical but the practical applications of a particular hypothesis. I don't think this is a possible way out, because, as I shall endeavour to show, the practical uses of a hypothesis are dependent on theoretical applications of it. Put rather more specifically my thesis will be this. A statistical hypothesis H is used in practice as follows. We first derive from H a prediction (theoretical use) and then base some action on this prediction. To see this let us consider a typical practical example. Suppose an industrialist is doubtful whether to buy machine A or machine B. He knows that the two machines are the same price, that their running cost per day is the same, and that the amount of material x which they process per day is normally distributed. The standard deviation of the distribution (σ say) is the same for both machines, but the means μ_A and μ_B say are different. We set up two statistical hypotheses, viz. $H_1 : \mu_A > \mu_B$ and $H_2 : \mu_A < \mu_B$. Some statistical method is used which results in the selection of H_1. Our problem is: assuming this selection to be correct, what action should the industrialist take?

At first the answer seems to be obvious and indeed to contradict our thesis. If $\mu_A > \mu_B$, then evidently machine A should be bought in preference to machine B. But let us now examine rather more closely why this conclusion follows. I claim that there is a suppressed premiss to the effect that if $\mu_A > \mu_B$, then machine A will on average process more material per day than machine B. But how is this suppressed premiss justified? If we have a falsifying rule, it is of course easy. The average amount of material processed by (average output of) machine A over n days is normally distributed with mean μ_A and standard deviation $\sigma/(n)^{1/2}$. Applying our falsifying rule we infer that the average output over a long period will be contained in a narrow interval centred on μ_A. The same applies for machine B. Therefore from hypothesis H_1, i.e. $\mu_A > \mu_B$, we can f-deduce that the average output over a long period of machine A will be greater than that of machine B. This reduces the practical application of H_1 to the required form. From H_1 we f-deduce a prediction about the average outputs of two machines. The industrialist can then base his action on this prediction.

Suppose however that we do not have a falsifying rule. Once again I cannot see how in these circumstances a connection could be established between $\mu_A > \mu_B$ and the average outputs of the two machines. Yet it seems to me necessary to establish such a connection in order to demonstrate the rationality of the decision. Without a falsifying rule, it is impossible to put a statistical hypothesis H to practical use in a way which can be rationally justified. I conclude that we cannot coherently adopt the Neyman–Pearson theory without a falsifying rule.

(iii) The Neyman–Pearson theory with a falsifying rule

Let us therefore turn to examining the second possibility and see whether a falsifying rule is implicit in the Neyman–Pearson theory, or at least can be formulated in terms of the Neyman–Pearson theory. To begin with we cannot adopt the following falsifying rule:

Suppose the range $R(\xi)$ of a random variable ξ with postulated distribution D can be partitioned into two disjoint sets A and C with $A \cup C = R(\xi)$, and suppose $\text{prob}(C) = k < k_0$ where k_0 is some suitable constant. Then if we observe $\xi \in C$, the hypothesis that ξ has distribution D must be taken as falsified.

This was rule R.4 which we discussed on pp. 176–9 above. We showed there that this rule enables us to make 'counter-intuitive' choices of critical region C, i.e. choices which involve regarding the hypothesis as falsified by evidence which intuitively we would consider as corroborating it. Now here we come to the essential difference between the Neyman–Pearson theory and our own view. We eliminated these counter-intuitive choices of C by requiring that C should have low l- as well as low k-value (and stipulating some further conditions besides). Neyman and Pearson eliminate these counter-intuitive choices by consideration of a set of alternative hypotheses. The 'counter-intuitive' choices of C are ruled out as having 'low power' relative to this set of alternatives. This point is made by Neyman and Pearson again in their first joint paper. They are discussing a sample Σ and the Hypothesis A that it is randomly drawn from the population Π. They say (1928, p. 9): 'But in the great majority of problems we cannot so isolate the relation of Σ to Π; we reject Hypothesis A not merely because Σ is of rare occurrence, but because there are other hypotheses as to the origin of Σ which it seems more reasonable to accept.' In favour of our own approach we can observe that it is simpler as it does not involve the consideration of alternative hypotheses, and, more important, it allows us to test statistical hypotheses 'in isolation', i.e. without formulating precise alternatives. This is something which we do want to do as I shall later show. The main conclusion I want to draw for the moment is that if we do formulate a falsifying rule within the Neyman–Pearson theory, it will have to involve some consideration of alternative hypotheses.

Bearing this in mind I suggest the following, which is the best I can do. I have formulated the rule only for simple hypotheses, and evidence consisting of n-tuples of observations, but it would be easy to generalize.

R.5 (*Neyman–Pearson Falsifying Rule*). Suppose we have a statistical hypothesis H which assigns a distribution to a random variable ξ. Suppose we have an evidence space E each point of which (**e** or **x**) consists of an n-tuple which possibly represents n independent values of ξ. Suppose further H is a point ω of a hypothesis space Ω, the other members of Ω being possible alternative hypotheses about the distribution of ξ. Then if C is a critical region of a test of H which has good Neyman–Pearson properties, e.g. is UMP or UMPU relative to Ω, we regard H as falsified if we observe $\mathbf{e} \in C$. Conversely the region $A = E - C$ is an 'acceptance region' for H.

This rule has, I believe, considerable plausibility when applied to the question of testing hypotheses. Its implausibility emerges when we attempt to use it to obtain statistical explanations or predictions. It will be remembered that we use a falsifying rule to obtain explanations and predictions in the following way. Suppose that from our hypothesis H we infer that a certain random variable ξ has a falsifiable distribution with acceptance region A. We then deduce in accordance with the falsifying rule (or f-deduce) that $\xi \in A$. If '$\xi \in A$' entails that some observed phenomenon occurs, we say that H explains that phenomenon. If $\xi \in A$ is an observable phenomenon which has not yet occurred, we can predict that $\xi \in A$ will occur. The case of prediction really covers that of testing, because in testing we predict that $\xi \in A$ and regard H as falsified if $\xi \notin A$. Let us now see how all this applies if we adopt R.5.

The trouble is that we cannot use a statistical hypothesis H to explain or predict an observable phenomenon unless we specify a set of hypotheses alternative to H. Only if we have such a set can we determine a critical region and hence an acceptance region. Moreover the predictions (regions of acceptance) we obtain will vary according to the set of alternatives we choose. This situation seems to me completely unacceptable. It is not *so* bad to test one hypothesis against a set of alternatives. I cannot however make any sense of the notion of obtaining an explanation with a given hypothesis only if we have a set of alternatives.

I will next show that the Neyman–Pearson procedure enables us to derive explanations and make predictions where intuitively we should not be able to do so. In our own theory, one can only make predictions regarding the value of a random variable ξ if ξ has a falsifiable distribution. If the distribution of ξ is non-falsifiable (say uniform) then no predictions are obtainable. This is fully in accord with intuition. If ξ has the distribution

$$f(x) = \tfrac{1}{2} \quad \text{for } -1 \leqslant x \leqslant 1$$
$$= 0 \quad \text{otherwise,}$$

then intuitively we cannot make any predictions about what value ξ will take in the closed interval $[-1, 1]$. Any value in that interval is, as far as our hypothesis goes, exactly similar to any other value. I will now show that the Neyman–Pearson procedure not only allows us to make predictions here, but it allows us to make different predictions according to the set of alternative hypotheses chosen.

Let us first consider the distribution in question as a special case of a series of distributions (cf. Diagram 1 below) defined by the equations

$$f(x, \alpha) = \tfrac{1}{2}\alpha \quad \text{for } -1 \leqslant x \leqslant -1 + \alpha$$
$$\text{and } 1 - \alpha \leqslant x \leqslant 1$$
$$= 0 \quad \text{otherwise}$$

where $0 < \alpha \leqslant 1$.

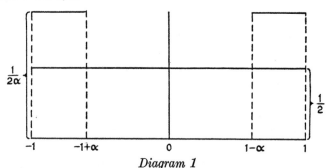

Diagram 1

Here the problem becomes that of testing $\alpha = 1$ against alternatives $0 < \alpha < 1$. In this case if we set

$$C = [-1, -1 + \epsilon] \cup [1 - \epsilon, 1]$$

C is an UMP test of size ϵ. Correspondingly we obtain the prediction $\xi \in (-1 + \epsilon, 1 - \epsilon)$.

However, we can also consider the distribution as a special case of a series of distributions (cf. Diagram 2 below) defined by the equations

$$f(x, \beta) = \tfrac{1}{2}\beta \qquad \text{for } -\beta \leqslant x \leqslant \beta$$
$$= 0 \qquad \text{otherwise}$$

where $0 < \beta \leqslant 1$.

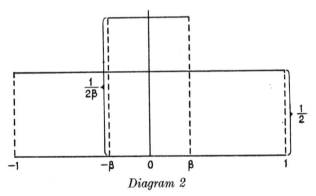

Diagram 2

Here the problem becomes that of testing $\beta = 1$ against the alternatives $0 < \beta < 1$. In this case if we set

$$C = [-\epsilon, \epsilon],$$

C is an UMP test of size ϵ. Correspondingly we obtain the different prediction $\xi \in [-1, -\epsilon) \cup (\epsilon, 1]$. I do not think this example is unfair to Neyman and Pearson because they themselves consider the case of rectangular distributions (but in n-dimensions) in their 1933a paper (pp. 159–62).

To sum up then. The Neyman–Pearson falsifying rule leads to the unacceptable notion that a statistical hypothesis H can only yield explanations of observable phenomena relative to a set of alternative hypotheses. Further, the rule enables us to obtain predictions in situations where, intuitively, no predictions should be obtainable. These predictions differ according to the set of alternative hypotheses adopted.

To this it could perhaps be replied: 'We agree that the Neyman–Pearson theory does not accord well with your idea of a falsifying rule. But this idea of a falsifying rule does not impress us very much, and we are not therefore disturbed by such a discord.' Certainly my arguments so far against the Neyman–Pearson theory depend on the notion of a falsifying rule, but I shall now try to give further arguments which are independent of this notion.

(iv) Two counter-examples from statistical practice

The idea of these arguments is to attack a principle which is shared by the Neyman–Pearson theory and several other approaches to statistics, for example decision theory. Thus although my main target is the Neyman–Pearson theory, my arguments could be used with slight modification against certain other views. We could call the principle in question the principle of alternative hypotheses. It states that when we are testing a given statistical hypothesis H, we can (and should) devise a set of alternatives to H and then represent the problem as that of testing H against these alternatives. Neyman and Pearson (1933b, p. 187) state the principle as follows:

> In any given problem it is assumed possible to define the class of admissible hypotheses $C(H)$, containing H_1, H_2,..., as alternatives to H_0.

Moreover, this principle is clearly necessary for the Neyman–Pearson theory because only by consideration of such alternatives can they rule out absurd choices of the critical region.

I am going to attack this principle of alternative hypotheses, but this does *not* mean that I am opposed lock, stock and barrel to the idea of devising alternative hypotheses when testing a given hypothesis. On the contrary, the history of science is full of examples of the fruitfulness of alternative hypotheses. I will content myself with mentioning what is perhaps the most famous example. In the early years of the sixteenth century, the Ptolemaic theory was a well worked out account of the universe which agreed with observation within the experimental error of the day, except perhaps for a few anomalies. Yet Copernicus devised an alternative hypothesis which agreed with observation as well as the Ptolemaic theory. This new hypothesis led to the production of new tests of the Ptolemaic account, and generally played a large part in stimulating the enormous scientific advances of the next hundred years. Granted then that alternative hypotheses can be of such value, why should I want to attack the principle of alternative hypotheses as it appears within the Neyman–Pearson theory?

There are really two reasons. First of all, although it is often desirable to devise alternatives when testing a given hypothesis, it is by no means necessary to do so. There are many situations where we want to test a given hypothesis 'in isolation', i.e. without formulating precise alternatives.[1] Indeed it is often the failure of such a test which elicits an alternative hypothesis. Suppose a hypothesis H is suggested, but as yet we have no precisely defined alternatives. Then on our account we can test out H. If it passes the tests, well and good. It can be provisionally accepted and used for some purpose. If on the other hand it fails the test, we will then try to devise a new hypothesis H' which avoids the refutation. In such cases the falsification of H provides the stimulus for devising an alternative hypothesis. Now if we stick to the Neyman–Pearson approach, the alternative hypothesis H' should have been devised before the very first test of H and that test should have been designed with the alternative in mind. The practising statistician can justly complain that this is too much to demand. He could point out that H might have been corroborated by the tests in which case the trouble and mental effort of devising an alternative would have been unnecessary. Further, he could argue that even if H

---

[1] This view is held by Fisher who writes (1956, p. 42): 'On the whole the ideas (a)...and (b) that the purpose of the test is to discriminate or "decide" between two or more hypotheses, have greatly obscured their understanding, when taken not as contingent possibilities but as elements essential to their logic. The appreciation of such more complex cases will be much aided by a clear view of the nature of a test of significance applied to a single hypothesis by a unique body of observations.' In a sense what follows can be considered as an attempt to support this opinion of Fisher's against that of Neyman and Pearson. Fisher's opinion has been supported recently by Barnard; cf. his contribution in L. J. Savage and others (1961).

is falsified, the nature of the falsification will give a clue as to what alternative might work better. It would be silly to start devising alternatives without this clue.

Now admittedly it is a very good thing if a scientist can devise a viable alternative H' to a hypothesis H, even when H has not yet been refuted. As I have just described, Copernicus devised an alternative astronomical theory even though the existing one (the Ptolemaic) was reasonably well corroborated. It is unreasonable, however, to demand that Copernicus' example be followed in every case. That would indeed be a counsel of perfection. In most cases it is only the refutation of a given hypothesis H which provides the stimulus for devising a new hypothesis H'. Consider then the schema. A hypothesis H is suggested. It is tested 'in isolation', and falsified. This falsification stimulates the production of an alternative hypothesis H' which works better. We shall show in a moment by means of examples that this schema frequently occurs in statistical investigations. However, it completely contradicts the model embodied in the Neyman–Pearson theory.

So far, when speaking of 'an alternative hypothesis' I have meant some hypothesis genuinely different from the one under test. But in practice Neyman and Pearson do not use 'alternative hypothesis' in such a sense, and this constitutes my second objection to their principle of alternative hypotheses. In practice the alternative hypotheses considered by Neyman and Pearson are nothing but the same hypothesis with different parameter values. Suppose, for example, that the hypothesis under test is that ξ is normal μ_0, σ_0, then the alternatives will be that ξ is normal with different μ, σ (or, in some cases, just with different μ). Thus the alternatives generally considered when the Neyman–Pearson theory is applied are merely trivial variants of the original hypothesis. But this is an intolerably narrow framework. We could (and should) consider a much wider variety of different alternatives. For example we might consider alternatives which assign a distribution to ξ of a different functional form. Again, we might reject the assumption that the sample $x_1, \ldots, x_n$ is produced by n *independent* repetitions of a random variable ξ and try instead a hypothesis involving dependence. We might even in some cases replace a statistical hypothesis by a complicated deterministic one. By restricting alternatives to such a narrow range, the Neyman–Pearson theory places blinkers on the statistician and discourages the imaginative invention of a genuinely different hypothesis of one of the types just mentioned. It must be remembered too that if a genuinely different hypothesis is proposed and corroborated, the fact that the original falsifying test was (say) UMP in a class of trivial variants ceases to have much significance.

To this argument a reply along the following lines will no doubt be made. 'These "academic" objections of yours are all very well. We fully admit that it would be nice to have a theory of testing which embodied all the alternatives of which you speak. But such a theory would be difficult, if not impossible, to construct. In such a situation the practical man must be content with the best we can do which is to consider only certain simple alternatives. Moreover, the Neyman–Pearson model embodying these simple alternatives finds frequent and useful application in statistical practice.' Against this I claim that the Neyman–Pearson model does not fit most statistical situations at all well and I shall try to show this by means of examples.

My first example of a statistical investigation which does not fit the Neyman–Pearson theory is taken, oddly enough, from Neyman himself (1952, pp. 33–7). The problem dealt with arose in the field of biology. An experimental field was divided into small squares, and counts of the larvae in each of these squares were made. The problem was to find the probability distribution of the number n of larvae in a square. The first hypothesis suggested was that this random variable had a Poisson distribution $p_n = \exp(-\lambda)\ \lambda^n/n!$ for some value of the parameter λ (i.e. a composite hypothesis). This was then tested by the χ^2-method. The possible results were divided into ten classes corresponding to 0, 1,..., 8, 9 or more, observed larvae. The number m_s ($s = 0, 1, ..., 9$) observed in each class was noted and the expected number m_s' was calculated given the hypothesis. For the purposes of this calculation the unknown parameter λ was estimated by the χ^2-minimum method. Finally, the value of the χ^2-statistic, namely $\sum_{s=0}^{9} (m_s - m_s')^2/m_s'$, was calculated. Under these circumstances it can be shown mathematically that the χ^2-statistic is approximately distributed in a χ^2-distribution with r-k-1 degrees of freedom where we employ

r classes and estimate k parameters from the sample. Here $r = 10$, $k = 1$. Thus we have 8 degrees of freedom. The value of the χ^2-statistic obtained was 46·8 and so we get a clear falsification. The details of the results are given in Neyman (1952, p. 33, Table III).

Let us pause to consider whether this test fits the general Neyman–Pearson theory. It is clear that it does not. We have a composite hypothesis that the distribution is Poisson for some value of the parameter λ. This hypothesis includes all possible parameter values and thus we cannot generate alternative hypotheses in the usual Neyman–Pearson fashion by varying one or more parameters. Of course this does not prevent us from obtaining alternative hypotheses in some other way. But does Neyman do so? Does he set up alternatives and try to find, say, an UMP test? Not at all. He accepts the usual χ^2-test without more ado. Let us next consider again possible 'counter-intuitive' choices of critical region, which I mentioned (p. 177). These consisted in taking for our critical region, not the tails of the distribution, but a narrow region in the 'head'. If this were done for the χ^2-dn in this case the result 46·8 would be a corroboration not a falsification. Naturally Neyman does not choose such a counter-intuitive critical region, but from the point of view of his own general theory he would be perfectly entitled to do so. On the Neyman–Pearson theory these 'counter-intuitive' critical regions are only ruled out because they have low power relative to some set of alternatives. Here we have no set of alternatives and thus the counter-intuitive regions become perfectly possible. On our own account, of course, such regions are ruled out because they have high l-values and this does not involve considering alternative hypotheses.

Let us now continue with our account of Neyman's investigation. The falsification of the Poisson hypothesis stimulated Neyman to produce a new hypothesis. This he arrived at by a brilliant heuristic argument which could well serve as a model of scientific reasoning. Neyman argues (1952, pp. 34–5) that we would expect a Poisson distribution if each larva is put on the field independently of the others. However, this is not what happens at all. A moth lays eggs at some point in the field which we can suppose to be randomly selected, i.e. the egg-laying points will follow a Poisson distribution. The moth lays a large number of eggs at once, and from these eggs hatch the larvae which then crawl slowly away from their birthplace in search of food. Naturally the number of eggs laid will vary from point to point, and also different numbers of larvae will die in each case. If the larvae crawl slowly, we will not expect them to be distributed randomly, i.e. according to the Poisson distribution. Rather we would expect them to be distributed in clumps of varying size about centres which are randomly distributed. Neyman proceeded to construct a mathematical distribution based on this mechanism. Its details are a little complicated and we will follow him in referring to it as a 'Type *A* distribution', remarking only that it depends on two parameters. Neyman then proceeded to test the Type *A* distribution using the χ^2-method just as before. The only difference was that two parameters were now estimated from the sample so that the χ^2-statistic had (approximately) the χ^2-distribution with 7 degrees of freedom. The result of the test this time was a definite corroboration, and consequently a triumphant vindication of Neyman's heuristic reasoning.

At this point it might be objected: 'You say that Neyman introduced a Type *A* distribution. Thus he did after all produce an alternative hypothesis as required by his general theory of testing.' I admit that Neyman did produce an alternative hypothesis, but only after the original Poisson hypothesis had been tested and falsified. Thus the alternative hypothesis was irrelevant to the first test of the Poisson hypothesis. This Neyman himself says (1952, p. 34):

In all cases, the first theoretical distribution tried was that of Poisson. It will be seen that the general character of the observed distribution is entirely different from that of Poisson. There seems no doubt that a very serious divergence exists between the phenomenon of distribution of larvae and the machinery assumed in the mathematical model. *When this circumstance was brought to my attention by Dr Beall, we set out to discover the reasons for the divergence.* [My italics]

In other words it was only *after* the first test that Neyman attempted to devise an alternative hypothesis. Indeed, as so often in science it was the *falsification* of a hypothesis which stimulated theoretical activity. Now of course the Type A hypothesis could have been devised before the first test. But as we pointed out above, it is unreasonable to demand scientific ingenuity of this level every time a hypothesis is tested. Moreover, had the Poisson distribution proved satisfactory, the mental effort needed to produce the Type A hypothesis would have been unnecessary.

A second point to notice concerns Neyman's handling of the second test, i.e. of the test of the Type A hypothesis. Now here a genuine alternative exists – namely the Poisson hypothesis. If Neyman had been true to his principles he should have tried to devise, say, an UMP test of the Type A hypothesis against the Poisson alternative. Of course he did not follow this course which would have involved him in difficult (perhaps impossibly difficult) mathematics, but instead used the standard χ^2-test. From our point of view he was eminently justified. The χ^2-test procedure falsifies one hypothesis and corroborates the other; it is thus genuinely crucial between the two hypotheses, and hence severe and to be commended. But is Neyman justified from the point of view of his own theory? Without a complicated mathematical investigation of the power properties of the χ^2-test, it is impossible to say.

For these reasons it cannot, I think, be denied that the piece of statistical reasoning as actually carried out by Neyman was not fitted into the Neyman–Pearson theory of testing. It might however be claimed that it could, as it were retrospectively, be fitted into the theory. That is to say we might, long after the event, propose some alternative hypotheses and show that the tests used were in some sense optimal against these alternatives. Indeed attempts have been made to fit the χ^2-test into the framework of the Neyman–Pearson theory. We must now examine whether these provide a solution to the present difficulty.

A typical such attempt is made by Lehmann (1959, pp. 303–6). Let us suppose we are testing the hypothesis H_0 that a certain real random variable ξ has a distribution $F(x)$. We will suppose the distribution completely specified so that the hypothesis is 'simple'. The test used is the ordinary χ^2-test with $k-1$ degrees of freedom. We therefore begin by partitioning the real line into k disjoint classes $C_1, \ldots, C_k$, and calculating the probabilities

p_{0i}, given H_0, of the result lying in C_i ($i = 1, \dots, k$). The χ^2-statistic is then calculated in the usual way. Lehmann proposes that we consider H_0 in the form H'_0:

$$\text{prob}\,(\xi \in C_i) = p_{0i} \qquad (i = 1, \dots, k)$$

and set up a set of alternatives H_t, say, defined by

$$\text{prob}\,(\xi \in C_i) = p_{ti} \qquad (i = 1, \dots, k)$$

where the p_{ti} are arbitrary non-negative constants subject to the conditions

$$p_{ti} \neq p_{0i} \qquad \left(i = 1, \dots, k \text{ and } \sum_{i=1}^{k} p_{ti} = 1\right).$$

He next shows that the χ^2-test has certain optimal properties relative to these alternatives. In fact these optimal properties are hardly very convincing. However I will not stress this point. Another difficulty is that in the present Neyman example we are considering a composite hypothesis in which the distribution of ξ contains certain arbitrary parameters. It is not clear that the present method will apply to this case. However, I propose to show that it is inadequate even in the simple case and thus *a fortiori* in the composite one.

My first objection is that, for the purposes of testing, H_0 is replaced by a hypothesis H'_0 which is not equivalent to it. In fact the assertion $\text{prob}\,(\xi \in C_i) = p_{0i}$ ($i = 1, \dots, k$) is compatible with many other distributions of ξ besides $F(x)$. Thus H'_0 is much weaker than H_0. This illegitimate replacement is not an accidental feature of the procedure. Suppose we retained H_0 in the form: ξ has the distribution $F(x)$. Then it would be natural to consider the alternatives: ξ has the distribution $G(x)$, where G is an arbitrary distribution different from F. However, relative to such a set of alternatives no power properties could be established. Thus H_0 has to be replaced by H'_0 which depends on a number of parameters. We can thus apparently generate satisfactory alternatives in the usual way by varying parameter values. However, these alternatives are really only plausible against a quite different hypothesis from the one actually being tested. A similar device is used by Neyman in constructing his 'smooth' test of goodness of fit

The next objection can be stated by introducing the notion of a 'serious' alternative. Let us say that H' is a serious alternative to H if we might actually adopt H' in the event of H being falsified. Now in my view it is of no value considering alternatives which are not serious in this sense; but the H_t as defined above are evidently not serious. Suppose, for example, that $F(x)$ is a continuous distribution. Then if we regarded H_0 as falsified we would probably try to replace it by a hypothesis H_1 which assigned a different continuous distribution to ξ (as in the present Neyman case). We would certainly not adopt a hypothesis of the form H_t in place of H_0. Anyone who doubts this should try to produce an example from statistical practice where a hypothesis of the form ξ, has continuous distribution F, has been falsified and actually replaced by a hypothesis of the form H_t. I feel certain that it will be impossible to do so.

The artificiality and inadequacy of the alternatives H_t is further illustrated by observing that they depend on inessential details of the actual testing procedure. Instead of the k classes $C_1, \ldots, C_k$ we could always employ $k-1$ or $k+1$ classes, or we could partition the real line into k classes in a different way. In all cases we would obtain a different set of alternatives H_t – illustrating the complete arbitrariness of the original set. Indeed the situation is usually made still worse by restricting the H_t to those of the form

$$\mathrm{prob}\,(\xi \in C_i) = p_{0i} + a_i/\sqrt{n} \qquad (i = 1, \ldots, k)$$

where the ai are fixed constants and n is the sample size. Why, one wonders, is $a_i/\sqrt{n}$ chosen rather than $a_i/(n)^{3/2}$ or any other function of n? The answer seems to be *not* that these alternatives are more realistic in any sense but that they give mathematically more interesting results! What is bad here, however, is not only the arbitrariness of the function of n but the fact that the alternatives should depend on n at all. This amounts to saying that if we draw a sample of 100, we should have one set of alternatives, whereas if we draw a sample of 150 we should have another set. Any genuine realistic alternative, however, would be proposed quite independently of the size of the sample used in the test.

Returning then to Neyman's piece of statistical inference, we can sum up as follows. The logical structure of the example is clear. A composite hypothesis H is proposed and tested 'in isolation', i.e. without formulating precise alternatives. It is falsified by the test, and a new hypothesis H' is produced by an ingenious heuristic argument. H' is in its turn tested, but this time it is corroborated. Neyman has provided us with a model example of scientific reasoning. It is strange that he does not see that it contradicts his general theory of testing which is expounded only twenty pages later in the same book.

My second example is of exactly the same type as the first. I only give it to show that cases of this type are not rare but on the contrary common in statistical reasoning. It is taken from Cramér (1945, p. 441). Cramér is considering the distribution of the breadths of beans of *Phaseolus vulgaris*. He has data for 12,000 of these beans. His first hypothesis is that the breadths are normally distributed for some mean μ and standard deviation σ. He applies the χ^2-test (estimating the two parameters) to this hypothesis, and a clear falsification results. Once again the standard Neyman–Pearson method of generating alternative hypotheses by varying parameters is not available, since all values of μ and σ are allowed as possibilities. Further, no other alternative hypotheses are suggested at this stage.

Cramér next argues thus. The normal distribution was suggested by the following considerations. It is reasonable to suppose that the deviations of the breadths of the beans from some mean value are caused by the operation of a large number of chance factors. Suppose the ith of these causes a small deviation ξ_i and there are n factors all told. Then the total deviation ξ is given by

$$\xi = \xi_1 + \cdots + \xi_n.$$

Now the central limit theorem yields that under certain very general conditions ξ will tend to a normal distribution as $n \to \infty$. But next suppose that n is not large enough for the normal approximation to apply. Can we get a better approximation to the distribution of sums like ξ? This mathematical question had been investigated by Cramér (1937), and it was natural for him to apply the results to the case in hand. In fact we do get a better approximation by adding to the normal frequency function successive terms of the Edgeworth series. Consequently Cramér modified his original hypothesis by adding the first terms of the Edgeworth series. He applied the χ^2-test, this time

15

estimating three parameters. The result was again a falsification. He then added the first *and second* terms of the Edgeworth series, applied the χ^2-test estimating four parameters, and obtained on this occasion a corroboration.

(v) Quantitative versus qualitative considerations

These then are my counter-examples drawn from statistical practice. Let us now pause for a moment to examine what objections could be raised to my arguments against the Neyman–Pearson theory. An objector might argue thus: 'You have shown that there are certain cases of statistical inference where the Neyman–Pearson model does not work very well, but of course there are many cases where it is entirely satisfactory. We agree that the Neyman–Pearson theory in general considers only simple alternatives obtained by varying parameter values. It would certainly be nice to have a theory which took into account more general alternatives – for example distributions of different functional form. However, such a theory would involve too many mathematical difficulties to be possible at the moment – though no doubt it will be possible in time. We are practical men and must content ourselves with the best that can be done at the moment. The trouble with your criticisms is that they are too negative. You complain about the deficiencies of the Neyman–Pearson theory without suggesting anything better to put in its place.'

My answer is this. Firstly, I deny that there are 'many cases where the Neyman–Pearson model works well'. I would claim on the contrary that it is hardly ever realistic. Secondly, to the criticism that I am attempting to demolish the Neyman–Pearson theory without putting anything in its place, my reply is that, if the proposed falsifying rule is accepted, the problem which the Neyman–Pearson theory sets out to solve no longer becomes so serious. It can, moreover, be dealt with by certain simple qualitative considerations without introducing a complicated mathematical theory. I will now elaborate this.

Let us suppose we have a statistical hypothesis H stating that a random variable ξ has a certain distribution for some values of a set of k parameters $(\mu_1, \ldots, \mu_k)$. Suppose further we have data consisting of a set $(x_1, \ldots, x_n)$ of n values of ξ. How can we test H? In order to do so we must find a statistic η, i.e. a function

$\eta(x_1, \ldots, x_n)$ of the sample which satisfies the following three conditions:

(1) It must be possible to calculate mathematically the distribution D of η given H, or at least to find a distribution D which is a good approximation to η's true distribution.

(2) D must be independent of the parameters $\mu_1, \ldots, \mu_k$.

(3) D must be a falsifiable distribution.

Our test then consists of regarding H as falsified if $\eta \in C$ where C is the critical region associated with D.

The standard statistical tests of course satisfy these conditions. In the χ^2-test we calculate the value of the χ^2-statistic, estimating the values of the k parameters by the χ^2-minimum method. Our mathematical theory shows that the χ^2-statistic if calculated in this way has approximately the χ^2-distribution with r-k-1 degrees of freedom where r is the number of classes employed. Further, the χ^2-distribution is falsifiable for a suitably large number of degrees of freedom. To take another example. Suppose we are using the t-test on the hypothesis that ξ has a normal distribution with zero mean and some standard deviation σ. We can then calculate that the t-statistic viz. $(n-1)^{1/2}(\bar{x}/s)$ where $\bar{x} = 1/n \, (x_1 + \cdots + x_n)$ and $s^2 = 1/n \sum_{i=1}^{n} (x_i - \bar{x})^2$ has the t-distribution with $n-1$ degrees of freedom. The t-distribution is of course independent of σ. Further, for suitably large n it is falsifiable.

My next point is that it is extremely difficult to find test-statistics satisfying conditions (1), (2) and (3). In fact only a handful of such statistics has been discovered for standard hypotheses, and we must consequently admire all the more the skill of those men who were able to find these statistics (i.e. K. Pearson, Student (W. S. Gosset), and R. A. Fisher). Neyman and Pearson represent the situation as one in which we have a very large number of possible tests and it is most important to select out of this number those which are best in some sense. This is not the case at all. Only a very few tests are available and it would require considerable ingenuity to devise more. Consequently, choosing a test will probably not be difficult and, if it is difficult, this is more likely to be because there is no test available than because there are a large number from which we must select one. We do not suffer from an *embarras de richesses* as far as tests are concerned.

It will no doubt be asked at this point: 'Why, if there really are only a few tests available, did Neyman and Pearson think there were so many?' The answer of course is: 'Because Neyman and Pearson required that the rejection class C should have a low k-value (low level of significance), but not that it should have a low l-value (low relative likelihood).' As we have pointed out several times, if we demand only a low k-value for C then a large number of choices of C become possible, many of them counter-intuitive. To eliminate these counter-intuitive choices Neyman and Pearson introduced their principle of alternative hypotheses, and required that we choose only those C which have high power relative to some set of alternatives. We, however, require that C should have not only a low k-value but also a low l-value (and that certain further conditions should be satisfied). This eliminates the counter-intuitive choices of C without the need for considering alternative hypotheses. Further it reduces drastically the number of possible tests. Tests can no longer be produced more or less mechanically as in the Neyman–Pearson theory. Ingenuity is necessary to devise a test. Consequently the problem of choosing between different possible tests is no longer such a serious one. We do not however wish to maintain that the problem disappears altogether. There may indeed be situations where a number of different statistical tests are genuinely available, and we want to choose one or two out of this number. What considerations should guide our choice? My answer is that certain simple qualitative considerations suffice for this purpose. There is no need to have a precise mathematical theory to determine our choice. Moreover, these qualitative considerations will not in general involve a consideration of alternative hypotheses. I will give an example of the kind of consideration I have in mind in the next section.

Against this it could be objected that quantitative considerations are preferable to qualitative ones, and hence the Neyman–Pearson theory is preferable to the view just expounded. However the Neyman–Pearson theory is just as dependent on qualitative considerations as I shall now show. This result follows in fact from a point made by Cox but used by him for a different purpose. I quote his remarks from L. J. Savage and others (1961, p. 84):

> Suppose that our simple hypothesis says that the density of the observations is $f_0(x)$, and that the test consists in calculating the function $t(x)$ and regarding large values of $t(x)$ as evidence against a null hypothesis. Suppose we consider the following family of hypotheses:
>
> $$f_\theta(x) = f_0(x)\, e^{\theta t(x)} / \int f_0(x)\, e^{\theta t(x)}\, dx.$$
>
> That is a family of hypotheses depending on the parameter θ; when $\theta = 0$ it reduces to the null hypothesis. Clearly the uniformly most powerful test of $\theta = 0$ is based on large values of t.

This shows that for null hypotheses of the type considered by Cox any test whatever is uniformly most powerful relative to some set of alternative hypotheses. Thus the property of being uniformly most powerful can only be significant if the set of alternatives introduced is in some sense realistic as opposed to arbitrary and artificial. But how do we decide that a set of alternatives is realistic? Only qualitative considerations will help us here.

More generally, on the view developed here we have to use qualitative considerations to choose between the various possible tests. This may seem a difficult task. In the Neyman-Pearson theory, however, we have to choose between the various possible sets of alternative hypotheses, which are often highly artificial (see pp. 213–14). We must apparently use qualitative considerations to decide between artificial and very artificial sets of alternatives. This is hardly an inviting task.

(vi) A reply to some objections of Neyman's
The time has now come to consider some of Neyman's objections.[1] They are expounded in Neyman (1952, pp. 43–54). Broadly speaking Neyman is attacking the view that it is possible to devise a good test of a hypothesis without taking into account alternative hypotheses. As he puts it himself (1952, p. 44): 'It is known that some statisticians are of the opinion

---

[1] I am grateful to Mr C. Howson for first drawing my attention to these.

that good tests can be devised by taking into consideration only the hypothesis tested. But my opinion is that this is impossible....' The view here under attack is certainly one to which I would subscribe. It seems to be perfectly possible in some circumstances to devise a good test of a statistical hypothesis without taking into account alternative hypotheses. Indeed I would cite Neyman's χ^2-test of the Poisson hypothesis as an example of this. It therefore becomes necessary to try and refute Neyman's arguments.

He proceeds by proving two mathematical results and then claiming that these results raise impossible difficulties for the position he is attacking. Like him I will begin by stating and proving these results – giving in fact rather simpler proofs based on a method of Cramér's. Throughout I will be concerned with the statistical hypothesis H that $x_1, \ldots, x_n$ are independent and normally distributed with zero mean and standard deviation σ. To test this we would customarily consider the t-statistic defined by $t = (n - 1)^{1/2} \bar{x}/s$ where

$$\bar{x} = 1/n(x_1 + \cdots + x_n)$$

$$s^2 = 1/n \sum_{i=1}^{n} (x_i - \bar{x})^2$$

However, Neyman considers the trivially different z-statistic $z = \bar{x}/s$, and we will follow him in this throughout the present section. The z-statistic given H has the distribution with frequency function

$$f(z) = 1/B[\tfrac{1}{2}(n - 1), \tfrac{1}{2}] (1 + z^2)^{-n/2}.$$

We shall call this the z-distribution with $n - 1$ degrees of freedom. Neyman's two results follow simply from the following lemma. In my statement of this lemma and the subsequent theorems I will take $z, x_1, \ldots, x_n$ to be as just defined.

LEMMA
If $(n)^{1/2}(\bar{x}') = \alpha_1 x_1 + \cdots + \alpha_n x_n$ where $\sum_{i=1}^{n} \alpha_i^2 = 1$ and $ns'^2 = \sum_{i=1}^{n} x_i^2 - n\bar{x}'^2$, then $\bar{x}'/s'$ has the z-dn with $n - 1$ d. of fr.

PROOF (cf. Cramér, 1945, pp. 379–82).
Consider $y_1, \ldots, y_n$ where

$$y_i = c_{i1} x_1 + c_{i2} x_2 + \cdots + c_{in} x_n$$

and (c_{ij}) is an orthogonal transformation, i.e. a rotation in n-space. The joint distribution of $y_1, \ldots, y_n$ is normal and we have

$$E(y_i) = 0$$

$$E(y_i y_k) = \sigma^2 \sum_{j=1}^{n} c_{ij} c_{kj} = \begin{cases} \sigma^2 & \text{for } i = k \\ 0 & \text{for } i \neq k \end{cases}$$

by definition of an orthogonal transformation. Thus the new variables y_i are uncorrelated. But they are normal. Therefore they are independent, i.e. $y_1, \ldots, y_n$ are (like $x_1, \ldots, x_n$) independent and normally distributed with mean zero and standard deviation σ. But now set $z_1 = (n)^{1/2} \bar{x}'$ so that

$$z_1 = \alpha_1 x_1 + \cdots + \alpha_n x_n \tag{1}$$

$\sum_{i=1}^{n} \alpha_i^2 = 1$. Therefore (1) is the first line of an orthogonal transformation. Extend this to a complete orthogonal transformation, and let the corresponding variables be $z_2, \ldots, z_n$. Then by the result just proved $z_1/(z_2^2 + \cdots + z_n^2)^{1/2}$ has the z-dn with $n - 1$ d. of fr. This is a standard result which is in fact usually used to introduce the z-dn (cf. Cramér, 1945, pp. 237–41). But since the transformation is orthogonal

$$x_1^2 + \cdots + x_n^2 = z_1^2 + \cdots + z_n^2$$

$$ns'^2 = \sum_{i=1}^{n} x_i^2 - n\bar{x}'^2 = \sum_{i=2}^{n} z_i^2$$

$$\bar{x}'/s' = z_1/(z_2^2 + \cdots + z_n^2)^{1/2}$$

and the result is proved. We can now present Neyman's two main results as the following two theorems.

THEOREM 1

In the situation under consideration we can find a statistic ζ of the r.v.'s $x_1, \ldots, x_n$ s.t.

(1) ζ like z has the z-dn with $n - 1$ d. of fr.

(2) $|z\zeta| \leqslant 1$.

PROOF

Set

$$(n)^{1/2} \bar{x}' = (x_1 - x_2)/(2)^{1/2}$$

$$ns'^2 = \sum_{i=1}^{n} x_i^2 - n\bar{x}'^2$$

$$\zeta = \bar{x}'/s'.$$

Then by lemma ζ has the z-dn with $n-1$ d. of fr. as required. The second property (cf. Neyman, 1952, p. 50) follows from some simple algebraic inequalities. We have for any real numbers a, b

$$(a \mp b)^2 \geqslant 0$$

$\therefore \qquad 2(a^2 + b^2) \geqslant a^2 + b^2 \pm 2ab = (a \pm b)^2$

But now

$$2n\bar{x}'^2 = (x_1 - x_2)^2 = ((x_1 - \bar{x}) - (x_2 - \bar{x}))^2$$
$$\leqslant 2\{(x_1 - \bar{x})^2 + (x_2 - \bar{x})^2\}$$
$$\leqslant 2 \sum_{i=1}^{n} (x_i - \bar{x})^2 = 2ns^2$$

$\therefore \qquad \bar{x}'^2 \leqslant s^2 \qquad\qquad (2)$

However

$$s'^2 + \bar{x}'^2 = s^2 + \bar{x}^2 = \left(\sum_{i=1}^{n} x_i^2 \right) \Big/ n$$

therefore (2) yields

$$\bar{x}^2 \leqslant s'^2 \qquad\qquad (3)$$

Therefore multiplying (2) and (3) and dividing by $s^2 s'^2$ we get

$$(x^2/s^2)(x'^2/s'^2) \leqslant 1$$

$\therefore \qquad |z\zeta| \leqslant 1 \qquad$ as required.

THEOREM 2

Suppose we have a sample $x'_1, \ldots, x'_n$ with at least one x'_i different from zero, then we can define a statistic ζ_0 say of the x_i's s.t.

(1) ζ_0 has the z-dn with $n-1$ d. of fr.
(2) ζ_0 takes the value $+\infty$ for the observed values.

PROOF

Set

$$\alpha_i = x'_i / (x'^2_1 + \cdots + x'^2_n)^{1/2} \qquad (i = 1, \ldots, n).$$

Define $\bar{x}'$ and s' as in above lemma and set $\zeta_0 = \bar{x}'/s'$. By the lemma ζ_0 has the z-dn. with $n-1$ d. of fr. However observed values of $(n)^{1/2}\bar{x}'$ is $(x'^2_1 + \cdots + x'^2_n)^{1/2}$ which is non-zero by hypothesis, and of ns' is zero. Thus the observed value of ζ_0 is infinite.

I must now explain the supposed paradoxical consequences

of these results. Let us take Theorem 1 first. ζ has the z-dn with $n - 1$ d. of fr. Therefore apparently we can use ζ to give a test of H. This will consist in rejecting H if the observed value of $|\zeta|$ is too large. Similarly we have of course the standard z-test which consists in rejecting H if $|z|$ is too large. But now by Theorem 1 $|z\zeta| \leqslant 1$. Thus if $|z|$ is large and the z-test gives a falsification, $|\zeta|$ will be small and the ζ-test will give a corroboration, and of course vice versa. As the two criteria will yield in this sense opposite conclusions, it looks as if it is necessary to choose between them. As Neyman puts it (1952, p. 45): 'Whenever one of these criteria has the most "improbable" values, thus "disproving" the hypothesis tested, the values of the other are just the most "probable" ones. This last circumstance will make it necessary to choose one of the criteria.'

But now let us examine how we might choose between z and ζ. If we consider only H, then apparently there is no difference between z and ζ as they both have the same distribution given H. The only way out seems to be an appeal to possible alternative hypotheses. In particular, if we adopt the Neyman–Pearson theory the solution is easy. We consider as alternative hypotheses H_μ that $x_1, \ldots, x_n$ are independent and normally distributed with some mean $\mu \neq 0$ and arbitrary standard deviation. Relative to H_μ, the x-test is UMPU but the ζ-test has no correspondingly good Neyman–Pearson property. Thus we would naturally select the z-test. But how could we make our choice without considering alternative hypotheses? Or, to put it another way, how could we eliminate the intuitively unsatisfactory ζ-test?

The difficulties raised by Theorem 2 are even easier to see. ζ^0 has the z-dn with $n - 1$ d. of fr. Thus we can apparently base a test on this statistic. But as ζ^0 has the value $+\infty$ for the given sample, the hypothesis will fail the ζ^0-test. Thus, whatever the sample, we can apparently obtain a test which will result in H being falsified, and the whole testing procedure is invalidated. Of course Neyman's solution would be to eliminate ζ^0 as not being, for example, an UMPU test.

These then are the difficulties which Neyman raises for those who reject the principle of alternative hypotheses as embodied in the Neyman–Pearson theory. I will now try to resolve these apparent paradoxes, starting with those generated by Theorem 1.

Neyman points out the peculiar relationship between the z-test and the ζ-test: namely that if H is falsified by the z-test it is corroborated by the ζ-test and vice versa. He concludes that this makes it necessary to choose between the two tests. As he puts it in a sentence already quoted (Neyman, 1952, p. 45), 'This last circumstance will make it necessary to choose one of the criteria.' Against this I maintain that the ζ-test and the z-test are both entirely valid tests. It would be possible with justice to apply either or both of them. The relationship between the two tests appears strange at first sight, but there is in fact nothing paradoxical about it. To establish this I propose to consider a hypothesis drawn from an unproblematic area of deterministic physics; to describe two tests T_1 and T_2 which everyone would accept as valid; and to show that T_1 and T_2 are related in the same way as Neyman's z-test and ζ-test.

The example I have in mind is none other than Galileo's law that falling bodies have in vacuo a constant acceleration of 981 cm/sec^2. We might be able to calculate from this law and certain assumed laws concerning the fracture of glass that if a steel ball of a certain size is dropped from a height h it will acquire a velocity sufficient to shatter and pass through a glass plate of thickness less than a, but that a glass plate of thickness greater than b will stop the ball without shattering. Now define test T_1 and T_2 as follows.

T_1: Drop a steel ball of the given size from height h onto a glass plate of thickness a_1 where $a_1 < a$. If the plate shatters and the ball continues its downward course Galileo's law is confirmed. If the plate stops the ball Galileo's law is falsified.

T_2: Drop a steel ball of the given size from height h onto a glass plate of thickness b_1 where $b_1 > b$. If the plate stops the ball Galileo's law is confirmed. If the ball shatters the plate and passes through, Galileo's law is falsified.

T_1 and T_2 are admittedly somewhat involved and impractical tests, but there is nothing wrong or paradoxical about them. We could with entire validity use either or both of them as tests of Galileo's law. But now observe that if T_1 falsifies Galileo's law, T_2 is bound to corroborate it, and vice versa. I conclude that there is nothing necessarily paradoxical about two tests being related in this way. It does not show that one of the tests is bad, or that we have to choose between them.

We see then that Neyman's claim that we must choose between the z-test and the ζ-test is not valid. Indeed it seems to me that we are quite entitled to use either or both of the two tests. This would be certainly true if we drew two samples of size n and applied the z-test using one sample and the ζ-test using the other. What is perhaps more doubtful is whether we can apply both tests to the same sample. In fact this raises the general problem of whether, given a single sample, we can use a variety of tests based on different statistics. If we do apply more than one test, we are, or at least might be, in a certain sense increasing the k-value employed. As we wish to keep the k-value below a certain level, this rather calls into question the whole procedure. However, in many situations it does seem reasonable to use a number of test statistics with only one sample. For example, in the coin-tossing case considered in Part II (pp. 124–6) we used a number of different statistics based on different gambling systems. As far as I can see no hard and fast rules can be laid down here, and it is a matter which must be left to common sense.

So there is no positive necessity for choosing between the z-test and the ζ-test. However we would probably wish to choose between them in practice, and in fact to select the z-test rather than the ζ-test. My next point is that a reason can be given for preferring the z-test to the ζ-test which does not involve alternative hypotheses. This refutes Neyman's claim that to make the choice we need to consider possible alternatives. What I have in mind is an application of the principle that we should prefer tests based on statistics which either measure quantities of practical interest or are closely related to such quantities. Now suppose, as would typically be the case in practice, that x_i measured the difference in yield of two grains A and B planted in the two halves of an experimental plot. Under these circumstances $\bar{x}$ measures for the sample the average increase (or decrease) in yield of A relative to B. Now this average increase or decrease is obviously of great practical importance because applied to the whole crop it will determine the gain or loss which will result if we use A rather than B. On the other hand $\bar{x}' = (x_1 - x_2)/(2)^{1/2}(n)^{1/2}$ does not give a measure for the sample of any quantity of practical importance. Applying the above principle we would prefer the z-test which is based on a normalized version of $\bar{x}$ to the ζ-test which is correspondingly related to $\bar{x}'$. This is an example of the 'qualitative considerations' mentioned on p. 218.

This ends my discussion of the difficulties raised by Theorem 1, but what of the problems created by Theorem 2? In fact when I was discussing randomness in Part II (pp. 121–4), I showed that if one is allowed to take account of the sample results, it is possible to design a test which refutes any given statistical hypothesis. Neyman's Theorem 2 shows the same result in a different way in this particular case. In Part II I suggested various ways in which this very real and perplexing difficulty might be surmounted, and I have nothing more to add now to that discussion. I conclude that Neyman's objections do not after all force us to adopt the principle of alternative hypotheses as it appears within the Neyman–Pearson theory, and that, although our approach is no doubt liable to many objections, it is at least not refuted by those which Neyman raises.

Explanation of Technical Terminology

Contents

In this book I am concerned solely with philosophical aspects of the theory of probability, and no mathematical points (as such) are discussed. Nonetheless I have found it convenient to make free use of mathematical *terminology*. I believe that an attempt to translate such terminology into ordinary language *in situ* would only result in a long-winded and obscure exposition. The mathematical terms are an indispensable aid to clarity and conciseness. Naturally, however, this decision may create difficulties for the philosopher interested in probability theory but unfamiliar with the standard mathematical terminology. The following appendix is an attempt to overcome these difficulties. I here explain in logical order all the technical terms which are used in the book, and include on p. 227 an alphabetical list of terms which gives for each the section of the appendix in which its meaning is defined. Any non-mathematician who reads through the appendix and then refers back to the technical terms as they appear in the text should be able to follow the argument of the book without any trouble. A word of warning is however needed. This appendix is not for mathematicians, whom I would rather refer to any standard account of probability theory, e.g. Kolmogorov's 1933 Monograph, or Doob's *Stochastic Processes* (I largely follow Doob as regards notation). I try in what follows to give a clear and not too inexact idea of the meaning of standard terms, but inevitably the account is somewhat loose and does not enter into all the subtleties and precise details which would be demanded by the mathematician.

1. The sample space (Ω)

In probability theory we are concerned with situations where some set of conditions is repeated, but where we do *not* always get the same result. Instead the result is one of a set of possible results. This set is called the *sample space*, and is denoted by Ω.

Example i. The conditions specify that we throw an ordinary die. The possible results here are 1, 2, 3, ..., 6. So $\Omega = \{1,2,3,...,6\}$.

Example ii. The conditions specify that we take an urn containing 200 white balls and 100 red balls. The balls are thoroughly mixed, one is drawn, its colour is observed, and it is then replaced. The possible results here are white and red. So $\Omega = \{\text{white,red}\}$.

Example iii. The die mentioned in Example i is thrown, not once but twice, and the results are noted. In this case the possible results are given by two numbers in a certain order or by an *ordered pair*. These possible ordered pairs are $\langle 1,1 \rangle$, $\langle 1,2 \rangle$,..., $\langle 1,6 \rangle$, $\langle 2,1 \rangle$,..., $\langle 6,6 \rangle$, and the set of them is Ω.

Example iv. Our conditions specify that a human being should be selected by some method from a certain population, and his or her height measured. Here the possible results are given by a figure in feet and inches, e.g. 5 ft 6 ins. Ω is the set of possible numbers of this sort between say 0 ft 0 ins and 9 ft 0 ins.

A particular result is denoted by ω. It is called a *member* or *point* of Ω, and we write 'ω is a member of Ω' in the form '$\omega \in \Omega$'. The points of Ω are called by Kolmogorov *elementary events*, and they specify completely a possible result of the underlying conditions, within the conceptual scheme which is being employed.

2. The union of two sets ($A \cup B$)

We have just seen that probability theory deals with sets of possible results which might be obtained when a set of conditions is repeated. It will therefore be helpful to explain some terminology used in the theory of sets. This will be done in Sections 2–4.

Given two sets of objects A and B, the union of A and B ($A \cup B$) is the set containing all the objects in either A or B or both.
  Example. If $A = \{1,2\}$ and $B = \{2,3\}$, then $A \cup B = \{1,2,3\}$.

3. The intersection of two sets ($A \cap B$)

Given two sets of objects A and B, the intersection of A and B ($A \cap B$) is the set containing just those objects in both A and B.
  Example. If $A = \{1,2\}$ and $B = \{2,3\}$, then $A \cap B = \{2\}$.

4. Disjoint sets

Two sets A and B are said to be disjoint if they have no members in common. Let us denote the *empty set* (i.e. the set with no members) by $\varnothing$. Then A and B are disjoint if, and only if, $A \cap B = \varnothing$.

5. The set of random events ($\mathfrak{F}$)

Let us now return to probability theory as such. As well as Ω, there is always specified a set of subsets of Ω (which always includes $\varnothing$ (the empty set) and Ω itself). These subsets are called *random events*. Suppose A is a random event, and a result ω occurs which is a member of A ($\omega \in A$). Then the random event A is said to have occurred. We will now give examples of random events in each of the cases considered under Section 1.

  Example i'. $A = \{2,4,6\}$. This is the random event 'an even number occurs'. If we get 2, 4 or 6 as the result of throwing the die, A is said to have occurred.
  Example ii'. $A = \{\text{white}, \text{red}\} = \Omega$. This is the random event 'a ball coloured white or red is drawn'. Obviously by the definition of the problem this event always occurs. It is the *certain event*.
  Example iii'. $A = \{\langle 1,6 \rangle, \langle 2,6 \rangle, \langle 3,6 \rangle, \langle 4,6 \rangle, \langle 5,6 \rangle, \langle 6,6 \rangle, \langle 6,5 \rangle, \langle 6,4 \rangle, \langle 6,3 \rangle, \langle 6,2 \rangle, \langle 6,1 \rangle\}$. This is the random event 'at least one 6 appears on the two throws of the die'. If we get a 6 on either (or both) throws, it is said to have occurred.

Example iv'. A = Set of all heights greater than 6 ft. This is the random event 'the height of the human being observed was more than 6 ft'.

In most simple examples the set $\mathfrak{F}$ of random events can be taken as the set of *all* subsets of Ω. However if Ω is a complicated set, $\mathfrak{F}$ has sometimes to be restricted. Very often $\mathfrak{F}$ contains only what are called *Borel* sets. In all cases $\mathfrak{F}$ has to be what is known as a *Borel field* or *σ-field*. However, these are mathematical subtleties which need not concern us.

6. Probability (p)
To each random event is assigned a probability. We will give typical examples which might appear in the cases considered under Section 1.

Example i". If the die is unbiased, we would assign to each of the possible results $1, 2, \ldots, 6$, the probability $1/6$.

Example ii". We would naturally suppose in this case that $\text{prob(white)} = 2/3$, $\text{prob(red)} = 1/3$.

Example iii". If each die is unbiased and the throws independent, we would assign to the random event 'at least one 6 appears on the two throws of the die' (see Section 5) $11/36$. (Proof Exercise).

Example iv". The probability of the event 'the human being observed was over 6 ft tall' would depend on the population and the means of selection. Even for a particular population and means of selection, its value could only be determined empirically.

7. A probability space ($\Omega, \mathfrak{F}, p$)
A probability space is just an ordered triplet consisting of a sample space Ω, the corresponding set of random events $\mathfrak{F}$, and the probabilities p assigned to each of the random events. Whenever probability theory is applied, it is assumed that we are dealing with a certain probability space.

8. A probability system ($\mathfrak{S}, \Omega, \mathfrak{F}, p$)
It will be remembered that the sample space Ω is the set of possible results of repeating a set of conditions. Call these conditions $\mathfrak{S}$. A probability system is then an ordered quadruplet consisting of $\mathfrak{S}$ followed by the corresponding $\Omega, \mathfrak{F}, p$. This concept has been introduced in the present book in order to investigate the relations between probability and experience.

9. The Kolmogorov axioms
These axioms were introduced by the Russian mathematician A. N. Kolmogorov in his 1933 monograph *Foundations of the Theory*

of Probability. They are accepted by the majority of mathematicians working on probability theory or statistics.

Axiom 1. For all $A \in \mathfrak{F}$, $0 \leqslant p(A) \leqslant 1$, and $p(\Omega) = 1$. (In words, the probability of any random event lies between 0 and 1 inclusive, and the probability of the certain event is 1.)

Axiom 2 (additivity). If A and B are two disjoint random events,

$$p(A \cup B) = p(A) + p(B).$$

It will be seen that all the probabilities introduced in Section 6 obey Axiom 1. Axiom 2 has also great intuitive plausibility. For example, if we are throwing an unbiased die, we would naturally assign the probability $1/3$ to the event $\{1,2\}$ and $2/3$ to the event $\{1,2,3,4\}$ but

$$\begin{aligned}
\text{Prob}(\{1,2\}) &= \text{Prob}(\{1\} \cup \{2\}) \\
&= \text{Prob}(\{1\}) + \text{Prob}(\{2\}) \\
&= 1/6 + 1/6 = 1/3. \\
\text{Prob}(\{1,2,3,4\}) &= \text{Prob}(\{1,2\} \cup \{3,4\}) \\
&= \text{Prob}(\{1,2\}) + \text{Prob}(\{3,4\}) \\
&= 1/3 + 1/3 = 2/3.
\end{aligned}$$

These calculations accord with Axiom 2. Axiom 2 is called the axiom of additivity for obvious reasons.

Normally mathematicians use a stronger form of Axiom 2 (Axiom 2′ say), which applies to a potentially infinite number of disjoint events. Naturally this is something of an abstraction and is really only introduced for mathematical convenience. Let $A_1, A_2, \ldots, A_i, \ldots$ be the infinite collection of events. We denote their union, i.e. the set whose members are in at least one of the A_i, by $\bigcup_i A_i$. We then have

Axiom 2′ (countable additivity). If $A_1, A_2, \ldots, A_i, \ldots$ are random events and A_i, A_j are disjoint for $i \neq j$ then

$$p\left(\bigcup_i A_i\right) = p(A_1) + p(A_2) + \cdots + p(A_i) + \cdots.$$

The phrase 'countable additivity' here means that the addition can be infinite. It should be remarked that although Axiom 2′ is generally accepted by mathematicians, it has recently been criticized by the notable Italian mathematician de Finetti who holds that one should use only 'additivity' (Axiom 2) and not 'countable additivity' (Axiom 2′).

Because probabilities are assigned to random events, i.e. sets, probability is sometimes referred to as a *set-function*. Because probabilities can be considered as measuring the chance of the random event occurring, probability is sometimes referred to as a *measure*.

16

Actually there is a general mathematical theory of the measures of sets of which probability theory is only a part. We can therefore formulate the Kolmogorov axioms (in their strong form) as follows: probability is a non-negative (i.e. positive or zero), countably additive, set-function defined for $A \in \mathfrak{F}$ and for which $p(\Omega) = 1$.

It is worth noting finally that 'probability' is not defined but appears as a primitive term in the Kolmogorov axioms. Much of the argument of the book hinges on this simple point.

10. Independence

Two random events A and B are said to be independent if and only if

$$p(A \cap B) = p(A)\,p(B).$$

This is sometimes called the product rule and corresponds, in a sense, to the axiom of additivity. It is worth noting however that the axiom of additivity applies to any two disjoint events A and B, while the product rule applies only to *independent* events (and indeed constitutes the definition of independence). It might also be asked how this formal definition of independence relates to the intuitive idea of independent events as those which 'do not affect or influence each other'. This is one of the topics discussed at some length in the book.

11. Cartesian product[1]

Suppose we have one set of repeatable conditions $\mathfrak{S}_1$ with possible outcomes Ω_1, and a second set $\mathfrak{S}_2$ with possible outcomes Ω_2. We can then consider the combined condition of repeating $\mathfrak{S}_1$ and then $\mathfrak{S}_2$ and noting the 'ordered pair' ($\langle \omega_1, \omega_2 \rangle$) of results. A special case of this was mentioned in Section 1, Example iii, and consisted of throwing a certain die not once but twice. The set of possible outcomes of the combined condition consists of all ordered pairs $\langle \omega_1, \omega_2 \rangle$ such that ω_1 is a member of Ω_1 and ω_2 of Ω_2. This set is known as the *cartesian product* of Ω_1 and Ω_2 and is denoted by $\Omega_1 \times \Omega_2$. Formally we have the definition

$$\Omega_1 \times \Omega_2 = \{\langle \omega_1, \omega_2 \rangle\} \quad \text{where} \quad \omega_1 \in \Omega_1 \quad \text{and} \quad \omega_2 \in \Omega_2.$$

In the same way we can consider the cartesian product of three sets $(\Omega_1 \times \Omega_2 \times \Omega_3)$ which is a set of ordered triplets etc.

12. Product measure[1]

Suppose now in the case discussed under Section 11 that A_1 is a subset of Ω_1 and A_2 of Ω_2, then $A_1 \times A_2$ is a subset of $\Omega_1 \times \Omega_2$. If

[1] The concepts introduced in Sections 11 and 12 are of less importance and may be omitted without much loss.

the two sets of conditions $\mathfrak{S}_1$ and $\mathfrak{S}_2$ are independent, we have by the product rule

$$p(A_1 \times A_2) = p(A_1)\,p(A_2).$$

In this way we can assign probabilities to subsets $A_1 \times A_2$ of $\Omega_1 \times \Omega_2$, or, to change terminology, we can say that we have assigned a measure to these subsets of $\Omega_1 \times \Omega_2$. (It will be remembered that we sometimes speak of probabilities as measures. See Section 9.) This measure is known as the *product measure*. As we are dealing with 2 sets (A_1, A_2) it can also be called a *2-fold* product measure. Similarly we can define 3-fold,..., n-fold product measures.

13. Random variable (r.v.) $(\xi, \eta, \zeta, \dots)$
Generally speaking the different possible results of repeating the underlying conditions are given by numbers. This was true of the die-throwing example (Section 1, Example i), where the possible results were 1, 2, 3,..., 6. It was also true of the case of selecting a human being from a population and measuring his or her height – for heights are evidently expressed by numbers. An exception however was the 'urn' example (Section 1, Example ii) where the possible results were colours (white or red) rather than numbers. Even in this case, however, we can express the result in numbers by using, for example, the code $0 = $ white, $1 = $ red. If the possible results of the underlying conditions are numbers, we can describe the situation using the concept of random variable.

A *random variable* is a variable whose value on a particular occasion is given as the result of repeating the underlying conditions. If we repeat the underlying conditions, the value obtained varies randomly – hence the name 'random variable'. We use the obvious abbreviation *r.v.* for random variable, and denote random variables by the small Greek letters ξ, η, ζ,....

Example. In the standard die-throwing case we have a random variable ξ whose value can be 1, 2, 3,..., 5 or 6.

Sometimes the possible results of the conditions are given by ordered pairs of numbers, e.g. $\langle 1, 2 \rangle$. In this case we speak of *2-dimensional (2-D)* random variables. Similarly we can introduce 3-dimensional (3-D),..., n-dimensional (n-D) random variables. A random variable whose value is an ordinary number can be called a 1-dimensional (1-D) random variable.

14. The range of a random variable $\xi(R(\xi))$
The set of possible values of a random variable ξ is called its *range*,

16*

and is denoted by $R(\xi)$. Obviously $R(\xi)$ is just the sample space, or in symbols

$$R(\xi) = \Omega.$$

15. The distribution (D) of a random variable

To each possible value of a random variable (and to the relevant sets of such values) we assign a probability. The totality of these probabilities is called the *distribution* (D) of the random variable. We can illustrate this in the simple urn example (Section 1, Example ii). Suppose we code the results by $0 =$ white, $1 =$ red. We can then introduce a r.v. ξ whose possible values are 0 and 1. The distribution of ξ is given by

$$\text{prob}\,(\xi = 0) = 2/3$$
$$\text{prob}\,(\xi = 1) = 1/3.$$

This distribution can be represented by the following diagram:

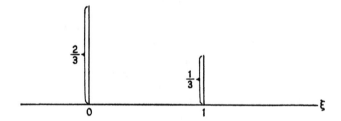

16. Discrete distributions

There are two important types of distribution, and we will now describe these in turn. We have first the case where the r.v. ξ can take on only a number of separate or *discrete* values. The possible values of ξ might be, for example, $a_1, a_2, \ldots, a_i, \ldots, a_n$ where for $i \neq j$, $a_i \neq a_j$. A particular case is the die example where $a_1 = 1$, $a_2 = 2, \ldots, a_6 = 6$. In this case we have a *discrete distribution* which can be specified by assigning a probability to the result that ξ takes the value a_i for each a_i. This can be done by giving a series of equations e.g.

$$\text{prob}\,(\xi = a_1) = p_1$$
$$\cdots$$
$$\text{prob}\,(\xi = a_i) = p_i$$
$$\cdots$$
$$\text{prob}\,(\xi = a_n) = p_n$$

where, of course, $p_1 + p_2 + \cdots + p_i + \cdots + p_n = 1$, and $p_i \geqslant 0$ for all i. Discrete distributions can be represented by diagrams of the type already introduced.

The height of each stroke is proportional to the probability of ξ taking the corresponding value.

17. Continuous distributions

Consider again the case of selecting a human being from a certain population and measuring his or her height (Section 1, Example iv). Suppose we express the result in decimals of a foot, e.g. 5·64 ft. Then in theory we could obtain any decimal number between certain limits. Here the possible values of the random variable are *continuous*, in the sense that there is a possible value arbitrarily close to any given value. In this case ξ is said to have a *continuous distribution*. It is easier in this case to look first at the graphical representation of a continuous distribution. In fact we use a curve of the following form:

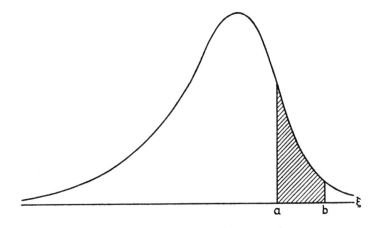

The meaning is this. The probability of getting a result lying between a and b, i.e. that $a < \xi < b$, is given by the area under the curve between a and b (shown shaded in the diagram). As anyone familiar with analytic geometry will know, the curve can be represented by an equation $y = f(x)$. $f(x)$ is known as the *frequency function* or *probability density function.* We denote the area under the curve between a and b by $\int_a^b f(x)\,dx$. So in symbols we have

$$\text{prob}\,(a < \xi < b) = \int_a^b f(x)\,dx.$$

Naturally the area under the whole curve must be 1. This is denoted in symbols by

$$\int_{-\infty}^{\infty} f(x)\,dx = 1.$$

18. The normal distribution

The most important continuous distribution is known as the normal distribution. Many random variables follow (at least approximately) a normal distribution. For example the heights of adult men in Britain are approximately normally distributed. The normal curve is the familiar symmetrical bell-shaped curve shown in the following diagram:

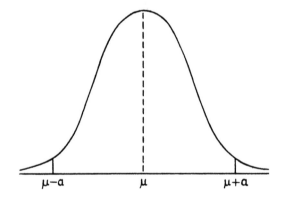

There is a very high chance of getting a result in a certain interval, $(\mu - a,\ \mu + a)$ say, round the mean value μ, and very little chance of getting a value which deviates from μ by more than a.

The equation of the normal distribution is given by

$$f(x) = (1/(2\pi)^{1/2}\,\sigma)\exp{-(x - \mu)^2/2\sigma^2}.$$

Here μ and σ are respectively the mean and standard deviation of the distribution. These terms will be explained in the next two sections. If a r.v. ξ has this distribution, it is said to be *normal* μ, σ.

19. Mean (μ)

Let us consider a random variable ξ which can take only the discrete values $a_1, a_2, \ldots, a_n$, and has the discrete distribution

$$\text{prob}\,(\xi = a_1) = p_1$$
$$\text{prob}\,(\xi = a_2) = p_2$$
$$\cdots$$
$$\text{prob}\,(\xi = a_n) = p_n$$

where $p_1 + p_2 + \cdots + p_n = 1$, and $p_i \geqslant 0$ for all i. We then define the *mean value* (or *mean*) μ of ξ by the equation

$$\mu = p_1 a_1 + p_2 a_2 + \cdots + p_n a_n.$$

Another name for the mean is the *expectation* or *expected value* $E(\xi)$ of the random variable ξ. This terminology is perhaps a little misleading because it suggests that μ, being the expected value of the r.v. ξ, should be the value of ξ which we expect. In fact, however, we cannot expect any particular value of the r.v. ξ whose value (by definition) varies randomly. μ is then better thought of as a kind of mean or average value of the r.v. ξ. To see this suppose each of the results $a_1, \ldots, a_n$ has the same probability $1/n$. Then

$$\mu = (a_1 + \cdots + a_n)/n.$$

In this case μ is the ordinary (arithmetic) average of the possible results. In general μ can be thought of as a kind of probabilistic average of the possible values of the r.v.

In the continuous case, the mean value of a random variable is similarly defined. An idea of its significance can be got by considering two normal distributions with the same standard deviation (σ), but with different means (μ_1 and μ_2).

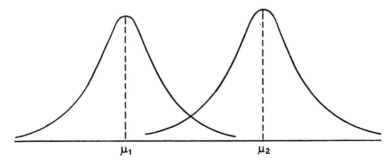

As can be seen, the distribution whose mean μ_2 is greater is displaced to the right. For this reason the mean is sometimes said to measure the *location* of the distribution.

Warning: For the normal distribution the mean value is also the value for which the curve becomes maximum. This is not true in general.

20. Standard deviation (σ)

Let us consider again the r.v. ξ taking discrete values $a_1, \ldots, a_n$ as defined in Section 19. Suppose ξ has mean value μ. Let us consider the square of the difference between μ and a_1 $(\mu - a_1)^2$. This is one measure of the degree to which a_1 differs from the mean value μ. If we now take the 'probabilistic average' of these differences we get

$$\sigma^2 = p_1(\mu - a_1)^2 + p_2(\mu - a_2)^2 + \cdots + p_n(\mu - a_n)^2.$$

σ^2 is known as the *variance* of the random variable. Its square root is the *standard deviation* of the r.v. So

$$\sigma = (p_1(\mu - a_1)^2 + \cdots + p_n(\mu - a_n)^2)^{1/2}.$$

Once again we can give an analogous definition in the case of continuous r.v.'s.

Intuitively the standard deviation is a measure of the degree to which the distribution of a random variable is spread out, or dispersed away from, its mean value. The larger the σ, the more the distribution is spread out. This is illustrated in the next diagram, which shows three normal curves with the same μ, but with increasing standard deviations $\sigma_1 < \sigma_2 < \sigma_3$.

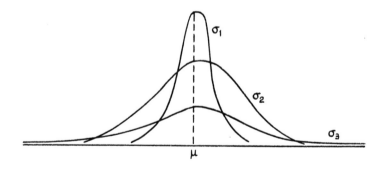

References

Works are referred to by the date of their first publication. If the edition from which quotations are taken differs significantly from the original one, it is also specified.

AYER, A. J. (1963) On the probability of particular events. The second of 'Two Notes on Probability', in *The Concept of a Person and Other Essays*, pp. 198–208. London: Macmillan.

BELL, A. E. (1947) *Christian Huygens and the Development of Science in the 17th Century*. London: Arnold.

BRAITHWAITE, R. B. (1953) *Scientific Explanation*. Cambridge: Cambridge University Press.

BRIDGMAN, P. W. (1927) *The Logic of Modern Physics*. New York: Collier-Macmillan. Paperback edition, 1960.

CANTELLI, F. P. (1935) Consideration sur la convergence dans le calcul des probabilités. *Ann. de l'institut Henri Poincaré*, V, pp. 1–50.

CARNAP, R. (1950) *Logical Foundations of Probability*. London: Routledge.

CHURCH, A. (1940) On the concept of a random sequence. *Bull. of the Am. Math. Soc.*, **46**, pp. 130–2.

COCHRAN, W. G. and COX, G. M. (1957) *Experimental Designs*. New York: Wiley.

COPELAND, A. H. (1928) Admissible numbers in the theory of probability. *Am. J. of Math.*, **50**, pp. 535–52.

CRAMÉR, H. (1937) Random variables and probability distributions. *Cambridge Tracts in Math.*, No. 36.

CRAMÉR, H. (1945) *Mathematical Methods of Statistics*. Princeton, N.J.: Princeton University Press.

DOOB, J. L. (1936) Note on probability. *Ann. Math.*, **37**, pp. 363–7.

DOOB, J. L. (1941) Symposium with von Mises on the foundations of probability and statistics. *Ann. of Math. Stat.*, **12**, pp. 206–17.

DOOB, J. L. (1953) *Stochastic Processes*. New York: Wiley.

DORLING, J. (1972) Bayesianism and the rationality of scientific inference. *Brit. J. for the Phil. of Sci.*, **23**, pp. 181–90.

DUGAS, R. (1958) *Mechanics in the Seventeenth Century*. Neuchatel: Griffon.

DUHEM, P. (1906) *The Aim and Structure of Physical Theory*. English paperback edition, London: Atheneum, 1954.

FELLER, W. (1957) *Introduction to Probability Theory and Its Applications*. 2nd edition. New York: Wiley.

FINETTI, B. DE (1937) Foresight: its logical laws, its subjective sources, in *Studies in Subjective Probability*, ed. by H. E. KYBURG JR. and H. E. SMOKLER. New York: Wiley, 1964.

FINETTI, B. DE (1970) *Teoria delle Probabilità*. Torino: Einaudi.

FISHER, R. A. (1956) *Statistical Methods and Scientific Inference*. Edinburgh: Oliver & Boyd.

FRY, T. C. (1928) *Probability and its Engineering Uses*. New York: Van Nostrand.

GALILEO (1938) *Two New Sciences*. Dover edition of English translation by HENRY CREW and ALFONSO DE SALVIO. New York: Dover, 1954.

GILLIES, D. A. (1972) The subjective theory of probability. *Brit. J. for the Phil. of Sci.*, **23**, pp. 138–57.

GNEDENKO, B. V. (1950) *The Theory of Probability*. English translation by B. D. SECKLER. New York: Chelsea, 1968.

GNEDENKO, B. V. and KOLMOGOROV, A. N. (1949) *Limit Distributions for Sums of Independent Random Variables*. Revised English translation by K. L. CHUNG. Reading, Mass.: Addison–Wesley, 1967.

HACKING, I. (1965) *Logic of Statistical Inference*. Cambridge: Cambridge University Press.

HACKING, I. (1967) Slightly more realistic personal probability. *Philosophy of Sci.*, **34**, No. 4, pp. 311–25.

HERIVEL, J. (1965) *The Background to Newton's 'Principia'*. London: Oxford University Press.

JEFFREYS, H. (1939) *The Theory of Probability*. London: Oxford University Press.

KENDALL, M. G. and BABINGTON SMITH, B. (1938) Randomness and random sampling numbers. *J. of the Roy. Stat. Soc.*, **101**, p. 147.

KENDALL, M. G. and BABINGTON SMITH, B. (1939a) Randomness and random sampling numbers. *J. of the Roy. Stat. Soc.* (Suppl.), **6**, p. 57.

KENDALL, M. G. and BABINGTON SMITH, B. (1939b) *Tables of Random Sampling Numbers*. Cambridge: Cambridge University Press.

KEYNES, J. M. (1921) *A Treatise on Probability*. London: Macmillan.

KHINTCHINE, A. J. (1952) Die Methode der willkurlichen Funktionen und der Kampf gegen den Idealismus in der Wahrscheinlichkeitsrechnung. *Sowjetwissenschaft Naturwissenschaftliche Abteilung*, 1954, pp. 261–73. German translation of a Russian original.

KOESTLER, A. (1959) *The Sleepwalkers.* London: Hutchinson. Penguin edition, 1968.

KOLMOGOROV, A. N. (1933) *Grundbegriffe der Wahrscheinlichkeits-rechnung.* English translation by N. MORRISON, *Foundations of the Theory of Probability.* New York: Chelsea, 1956.

LAKATOS, I. (1968) Changes in the problem of inductive logic, in I. LAKATOS, ed., *The Problem of Inductive Logic,* pp. 315–417. Amsterdam: North-Holland.

LEHMANN, E. L. (1959) *Testing Statistical Hypotheses.* New York: Wiley.

LENIN, V. I. (1908) *Materialism and Empirio-Criticism: Critical Notes on a Reactionary Philosophy.* Moscow: Progress Publishers, 1970.

LOVELAND, D. W. (1966) A new interpretation of the von Mises' concept of random sequence. *Zeit. Math. Logik und Grundlagen der Math.,* **12**, pp. 277–94.

MACH, E. (1883) *The Science of Mechanics: A Critical and Historical Account of its Development.* 6th edition. LaSalle, Ill.: Open Court Publishing Co., 1960.

MARTIN-LÖF, PER (1966) The definition of random sequences. *Inf. and Control,* **9**, pp. 602–19.

MISES, R. VON (1919) Grundlagen der Wahrscheinlichkeitsrechnung, in VON MISES' *Selecta II,* pp. 57–106. Providence, Rhode Island: American Mathematical Society, 1964.

MISES, R. VON (1928) *Probability, Statistics and Truth.* 2nd revised English edition. London: Allen & Unwin, 1951.

MISES, R. VON (1931) *Wahrscheinlichkeitsrechnung.* Leipzig und Wien: F. Deuticke.

MISES, R. VON (1938) Ernst Mach und die empiristische Wissen-schaftsauffassung, in *Selecta II,* pp. 495–523.

MISES, R. VON (1940) Scientific conception of the world: On a new textbook of positivism, in *Selecta II,* pp. 524–9.

MISES, R. VON (1941) On the foundations of probability and statistics, in *Selecta II,* pp. 340–55.

MISES, R. VON (1952) Sur les fondéments de calcul des probabilités. *Theorie des Probabilités, Exposés sur ses fondements et ses applica-tions.* Paris: Gauthier Villars. Pp. 17–29.

MISES, R. VON (1963) *Mathematical Theory of Probability and Statistics.* (Von Mises died in 1953. This volume was edited from his posthumous papers by his widow, Hilda Geiringer.) New York and London: Academic Press.

NEWTON, I. (1687) *Principia.* Cajori's edition of Motte's English translation of the 3rd edition. Berkeley, Calif.: University of California Press, 1934.

NEWTON, I. (1728) *The System of the World*. In Cajori's edition of the *Principia*, cited above.

NEYMAN, J. (1952) *Lectures and Conferences on Mathematical Statistics and Probability*. 2nd edition, revised and enlarged, Washington: Graduate School of U.S. Dept. of Agriculture.

NEYMAN, J. and PEARSON, E. S. (1928) On the use and interpretation of certain test criteria for purposes of statistical inference, in *Joint Statistical Papers of J. Neyman and E. S. Pearson*, pp. 1–98. Cambridge: Cambridge University Press, 1967.

NEYMAN, J. and PEARSON, E. S. (1933a) On the problem of the most efficient tests of statistical hypotheses, in *Joint Statistical Papers*, pp. 140–85. Cited above.

NEYMAN, J. and PEARSON, E. S. (1933b) The testing of statistical hypotheses in relation to probabilities a priori, in *Joint Statistical Papers*, pp. 186–202. Cited above.

PEIRCE, C. S. (1910) Notes on the doctrine of chances. Reprinted in *Essays in the Philosophy of Science*. New York: American Heritage Publishing Co., 1957.

POINCARÉ, H. (1902) *Science et Hypothèse*. English translation by F. MAITLAND, *Science and Hypothesis*. New York: Dover, 1952.

POPPER, K. R. (1934) *Logik der Forschung*. English edition, *The Logic of Scientific Discovery*. Translated by the author with the assistance of DR J. FREED and L. FREED. London: Hutchinson, 1959.

POPPER, K. R. (1957a) Probability magic or knowledge out of ignorance. *Dialectica*, **11**, No. 3/4, pp. 354–74.

POPPER, K. R. (1957b) The aim of science. *Ratio*, **I**, No. 1, pp. 24–35.

POPPER, K. R. (1959) The propensity interpretation of probability. *Brit. J. for the Phil. of Sci.*, **10**, pp. 25–42.

POPPER, K. R. (1963) *Conjectures and Refutations*. London: Routledge.

RAMSEY, F. P. (1926) in *Truth and Probability in the Foundations of Mathematics*, ed. by R. B. BRAITHWAITE, pp. 156–98. London: Routledge.

RÉNYI, A. (1955) On a new axiomatic theory of probability. *Acta Math. Acad. Sci. Hung.*, **6**, pp. 285–335.

ROLLER, D. (1957) The early developments of the concepts of temperature and heat, in *Harvard Case Histories in Experimental Science*, ed. by J. B. CONANT, Vol. I, pp. 119–214. Cambridge, Mass.: Harvard University Press.

RUSSELL, B. (1914) *Our Knowledge of the External World*. London: Allen & Unwin.

RUTHERFORD, D. E. (1951) *Classical Mechanics*. Edinburgh: Oliver & Boyd.

SAVAGE, L. J. (1954) *The Foundations of Statistics.* New York: Wiley.

SAVAGE, L. J. and others (1961) *The Foundations of Statistical Inference.* London: Methuen.

SUPPES, P. (1966) *Probabilistic Inference and the Concept of Total Evidence in Aspects of Inductive Logic*, ed. by J. HINTIKKA and P. SUPPES. Amsterdam: North-Holland.

TODHUNTER, I. (1865) *A History of the Mathematical Theory of Probability from the Time of Pascal to that of Laplace.* New York: Chelsea, 1965.

TORNIER, E. (1933) Grundlagen der Wahrscheinlichkeitsrechnung. *Acta Math.*, **60**, pp. 239–380.

USPENSKY, J. (1937) *Introduction to Mathematical Probability.* New York: McGraw-Hill.

WALD, A. (1937) Die Widerspruchsfreiheit des Kollektivbegriffes. *Ergeb. eines Math. Kolloq.*, **8**, pp. 38–72.

WALD, A. (1938) Die Widerspruchsfreiheit des Kollektivbegriffes in *Wald: Selected Papers in Statistics and Probability*, pp. 25–41. New York: McGraw-Hill, 1955.

WITTGENSTEIN, L. (1953) *Philosophical Investigations.* Oxford: Blackwell.

Index to Appendix

General Index

acceptance of theories, 65–9
acceptance region, 170, 172f, 192, 203
additivity (finite versus countable), 25
d'Alembert, 155, 164–7
alternative hypotheses, 178, 196f, 202f, 219f
attribute space, 3, 84
Atwood's machine, 40
axiom of convergence, 4, 6, 147–8
axiom of independent repetitions, 33, 81, 99, 103–4, 110, 112, 119, 132, 150–4
axiom of randomness, 6, 82, 88, 147, 152
Ayer, A. J., 154, 156

Babington Smith, B., 124, 128–30
Barnard, G., 207
Bayes' theorem, 17, 18
Bayesian conditionalization, 19, 21–3
Bell, A. E., 48
Bernoulli, J., 11
Bertrand, J., 11
Bohr, N., 72
book paradox, 11
Borel, E., 11, 103
Braithwaite, R. B., 132, 144, 180–6
Bridgman, P. W., 42–5
Buffon, 126–7, 165–7

Cantelli, F. P., 83
Carnap, R., 28–9
centrifugal force, 50
Cesaro, 118
Chebichev, P. L., 181–2, 184

Church, A., 84–7, 122
Cochran, W. G., 122–3
coherence condition, 15, 17
collective, 3f, 82–6, 127, 143, 150
combining repeatable conditions, 96f
comets, Newton's theory of, 68
composite hypothesis, 196f, 209f, 213
conceptual innovation
  problem of, 47
  theory of, 63
conditional bet, 16, 20, 21
conditional probability, 8
Condorcet, 6
confirmation, 28–31
Constantinople, 155
Copeland, A. H., 84, 130–1
Copernicus, N., 207–8
correspondence, generalized principle of, 71–3, 90, 139, 187
corroboration, 28–31
Cournot, A. A., 2
Cox, G. M., 122–3, 218–19
Cramér, H., 25–7, 80, 87, 93, 102, 109, 215, 220–1
critical region, 170, 172f, 195f

deductive model of explanation, 69–71, 172
definitional thesis, 6, 37–8
degree of belief, 15–16, 20–1, 24, 28
degree of partial entailment, 8, 24
degree of rational belief, 9, 14, 15, 24, 28, 30
depth of a theory, 69–71
Descartes, R., 48
determinism, 133